Advance Praise for *Your Fertility. Your Family.*

"If you're down a blog rabbit hole of reasons why you aren't pregnant and feeling frustrated, please take a deep breath. Then make an appointment with CCRM in your nearest city. With some of the highest success rates in the country and exceptional care teams, you will feel control over your path to parenthood. Trust me, wanting the best for your baby starts before they are born, CCRM is the best!"

–Giuliana Rancic, Mother, E! Host,
Entrepreneur, Designer, Author

YOUR FERTILITY.
YOUR FAMILY.

WILLIAM SCHOOLCRAFT, M.D., HCLD
AND THE STAFF AT CCRM FERTILITY

A SAVIO REPUBLIC BOOK
An Imprint of Post Hill Press

ISBN: 978-1-64293-161-7
ISBN (eBook): 978-1-64293-162-4

Your Fertility. Your Family.:
The Many Roads to Conception
© 2019 by William Schoolcraft, M.D., HCLD
All Rights Reserved

Cover Design by Cody Corcoran

posthillpress.com
New York • Nashville

Published in the United States of America

TABLE OF CONTENTS

INTRODUCTION

IT GETS BETTER

When the CCRM staff and I wrote our first patient's guide to infertility, *If at first you don't conceive*, we felt compelled to provide simple, easy-to-understand information for those new to the infertility world so they could make wise decisions about their personal treatment path. The book outlined the initial steps for pursuing testing, diagnosis, and treatment for a range of fertility challenges.

Those who read the book said they found it very helpful as a beginner's guide into the often complex world of fertility treatment. Five years later, I realized that while our first book provided a solid starting point for patients, much of the material was rapidly becoming out of date due to scientific and medical advances.

The world of infertility treatments has changed substantially in less than a decade. Much like computer technology, software, cell phones, and even the music industry, the field of infertility science is transforming at a stunning rate. What was considered standard care and treatment only six or seven years ago is now thought of as "old school" and passé. In *Your Fertility. Your Family.*, we provide the latest treatment template used to diagnose and overcome fertility challenges. In addition, our growing CCRM team addresses new treatments and options that have emerged and are expected to emerge over the next decade.

When I describe our team as *growing*, I mean it quite literally. When we wrote the first book, CCRM, based outside Denver in Lone Tree, Colorado, had affiliate clinics in Houston and Michigan. Since then, we've added clinics in Toronto, Minneapolis, Atlanta, Newport Beach, Northern Virginia, Boston, and New York, and we are not done expanding our brand by any means. Several brilliant team members from our clinics have contributed chapters to this book, and for that we are very grateful.

In the following chapters, we offer up to date information on paths to conception. These include recently developed methods for maturing a woman's eggs in vitro, along with new data comparing frozen versus fresh embryo transfers. We will also look at the increasing benefits of comprehensive chromosomal screening.

Another major development is the growing trend of fertility preservation among career women, especially those reaching their mid-thirties and forties. Over the past five years, more women have been delaying childbearing and choosing to freeze their eggs. This book examines the emerging opportunities in elective fertility preservation, as well as egg freezing for those diagnosed with cancer or other diseases which lead to a loss of ovarian function.

As women turn more and more often to alternative forms of reproduction, this book also provides the latest information on the use of gestational carriers and egg donation. Single embryo transfer, for example, also has become much more common. Ten years ago, it was used for less than 10 percent of all IVF cycles in the United States, but now it is the option chosen for at least half of all transfers. We will note that this trend has evolved to provide the best opportunities for success, not only in achieving pregnancy, but having a safe, full-term, uncomplicated pregnancy for mother and baby.

Furthermore, we will provide information on new methods of diagnostic evaluation, including assessing one's fertility for the future and learning how long your fertility may be viable. Alternative treatments have always been of great interest to infertile patients and we'll look at those from the holistic mind-body perspective. Finally, we will provide information about the future of reproductive science and infertility treatments as we look forward using the lens of cutting edge developments and research. While the last ten years have been mindboggling with their advancements, we expect a similar rate of technology change over the next decade.

As a special bonus, this new book has a guest celebrity co-author, the television personality and entrepreneur, Giuliana Rancic, who has been a patient at CCRM and a champion of our work. This determined and courageous woman will share her own inspiring, yet complex journey, to motherhood, which took her through periods of great frustration and some of life's toughest challenges, but ultimately resulted in joyous success.

When she came to us, Giuliana was planning another IVF cycle, having failed previously at other centers. During our extensive testing, it was discovered that she had breast cancer, which delayed and complicated her fertility treatment, but ultimately did not prevent her from having a healthy son, and may well have saved her life.

Giuliana is a wonderful advocate and role model for patients with fertility challenges and for those with breast cancer. I have no doubt that her chapter will inspire readers.

I want to thank you in advance for purchasing our book and I hope you find it informative, enlightening, and inspiring. If you are facing your own challenges with infertility, please let us know if we can help you in any other way.

CHAPTER ONE

GIULIANA RANCIC'S JOURNEY TO MOTHERHOOD

Most people know me as the anchor of *E! News*, and as a co-star of *Fashion Police,* as well as other endeavors in the entertainment and fashion world. Yet, for many years I had a dual life. Sure, I was the upbeat and cheerful television personality, but I also was a married woman struggling to overcome infertility as well as life-threatening health issues.

One of those lives was very glamorous and still feels like the dream. The other was more like warfare. If you've ever faced challenges to your fertility, you know what I mean. In writing this, my hope is you will benefit from my experiences and encouragement. I am a seasoned veteran of the infertility battlefield, and I wanted to share my story because I am the poster child for overcoming infertility challenges.

Like many women I had assumed that I would have no trouble having children when I was ready. I was so wrong about that.

Now one of my life's missions is to get the word to young women before it is too late for them. Don't take your fertility for granted. I encourage all women in their late twenties and early thirties who want to have children one day to have a fertility assessment done at a reputable clinic so that you at least know where you stand.

I fought infertility on many fronts. I had a very difficult, but ultimately successful experience after overcoming some of the most common as well as the most difficult challenges of motherhood. One of them was cancer.

On this front too, I serve as a very good example of a bad example because I believed that there was no need to have a mammogram until I reached the age of forty. There was no history of breast cancer in my family and I was not a smoker, so the thought of having a mammogram before my fortieth birthday had never crossed my mind.

Doctor Schoolcraft is an advocate of mammograms for any woman over the age of thirty-five considering pregnancy, and CCRM makes this part of their protocol. If there is a known family history of breast cancer, and you are under the age of thirty-five, your physician may discuss the importance of having a mammogram prior to starting treatment. Had it not been for his insistence for me to have a mammogram, I might have died in my forties.

For that reason, it is not a stretch to say my fight to overcome infertility actually saved my life. And so, I stand with Doctor Schoolcraft and many others in encouraging all women to have regular mammograms and to be vigilant in examining their breasts for signs of cancer—even if there is no history of it in their families.

❁ *HOLDING ONTO HOPE*

Now, for the good news, which I hope you will take to heart. After many disappointments, as well as considerable pain and suffering, my husband, Bill, and I were able to conceive and have a child with the help of Doctor Schoolcraft's team, the wonders of advanced reproductive science, and a very special surrogate mother.

As I'm sure you will understand, Bill and I wanted to have at least one more child as a sibling to our son, but those efforts ultimately failed. I haven't given up hope for having another child with our shared DNA, because reproductive science is advancing rapidly. Please keep that in mind as you read this book. They are already doing amazing things like uterine transplants that will likely become common in the very near future. Please, don't give up. Keep fighting. Take it from me, there may be times when you think you've exhausted every possibility, but that can change quickly in today's world.

I had a long trip on the infamous roller coaster ride experienced by so many other women with infertility issues. I am no braver, no stronger, and certainly no more worthy of motherhood than any others. It's also true that my journey, while quite difficult and exhausting, was not all that unusual.

I was fortunate to have a compassionate husband, sufficient resources, and a resilient spirit rooted in faith. Bill has shared this crazy ride with me and supported me throughout. There aren't many husbands who also would have agreed to allow the public into such a private part of life.

Our struggle with infertility became a major part of the reality television show Giuliana & Bill, which began two years after our marriage and continued for seven seasons. I have also talked and written about it extensively in magazine interviews and in my books, I Do, Now What? and, more recently in my memoir, Going Off Script.

My goal in sharing my story in this book is to provide Doctor Schoolcraft with an example that other women can identify with and, hopefully, draw upon as a source of hope. If nothing else, you will see that you are not alone in your feelings. There are many, many women like you and me engaged in this struggle.

One more thing I want you to know. Every time I look into the eyes of our son, Duke, every time he hugs me ,and every night when I tuck him into bed, I am reassured that all of the anguish and pain was worth it.

❀ *MY PATH TO MOTHERHOOD*

I have no regrets about being a career-oriented young woman. Marriage and family were not priorities for me in my twenties and early thirties. I chased my career dreams in television broadcasting instead of chasing guys. I thought the right man would come around one day, and when that day came, I'd start thinking about having children.

My career dreams came true and the right guy did show up while I was at work. Bill Rancic came into my broadcasting world after gaining fame as the first winner of *The Apprentice* reality television show. The handsome Chicago real estate developer and entrepreneur won the show and then he won my heart.

I was the co-anchor for *E! News* and I interviewed this genuine and caring guy after Donald Trump hired him to oversee construction of Trump Tower in Chicago. He asked me for a date after the interview and we quickly fell in love, just as I'd always dreamed. I was thirty-two when we married in 2007.

After our wedding and honeymoon, I still thought there was plenty of time to settle in together and enjoy each other's company before we tried to have children. We wanted at least four. We each come from large families. I have forty first cousins. We always assumed we would have many children without any problem. I was shocked to learn that, at thirty-five, I was pushing my luck.

We had tried for about a year to make a baby. When it didn't happen in the first six months we got a bit more serious. We eased back on our schedules, slowed down our lives and tried to reduce stress, which we'd heard could affect our fertility.

❀ *IF AT FIRST YOU DON'T CONCEIVE*

When that didn't work, we consulted with my gynecologist in Chicago. He ordered some tests for both of us. The results showed that my egg quality was not all that good and that my uterus was ":misaligned." These didn't appear to be major issues. My colon was pushing my uterus out of alignment so a colonic fixed that. I took hormone shots to produce more follicles and eggs and we followed the doctor's instructions on the optimal timing to have sex.

After two months, we upped our game by trying intrauterine inseminations (IUI) to inject Bill's sperm directly into my uterus. The goal was to increase the number of sperm that reached my fallopian tubes to increase the chances of fertilization. I also increased my hormone injections to increase the number of follicles, as well as getting shots to induce ovulation.

None of this sounds very sexy, does it? It wasn't. Taking all of those hormones made me miserable because I seemed to suffer every possible nasty side effect. These included bloating, cramps, breast pain, and mild mood swings. If I'd become pregnant then and there, none of the misery would have mattered, of course. But no baby resulted.

Bill and I are strivers. We've always been dogged in pursuing our individual and shared goals. We're used to winning at whatever we pour our heart and soul into. We didn't take losing well.

Infertility was winning, and we were frustrated and more than a little ticked off. On the other hand, we weren't about to give up. We share the scrappy gene. So, we didn't quit. We doubled our efforts and our research.

❁ *TURNING TO THE INFERTILITY SPECIALISTS*

As part of my job on the entertainment news beat, I was always doing red carpet interviews with actresses who'd become pregnant in their thirties and forties and that was one reason I hadn't been all that concerned about getting pregnant myself. Once I delved into the reality of infertility and reproductive science I realized that some of those sexy ladies and their sexy spouses had achieved pregnancy only with the help of fertility specialists.

My gynecologist agreed that Bill and I should follow that path and seek a higher level of expertise with a top reproductive endocrinologist in Chicago. As we expected, the fertility specialist recommended that we try in vitro fertilization, which had progressed dramatically since the 1970s when the first children produced in this manner were known as "test tube babies."

IVF is now widely accepted as offering the best chances for a healthy pregnancy, but that doesn't mean it's easy, pain-free, or inexpensive. We learned that although in vitro fertilization had become the option of choice for most infertile couples, it remained a complex, invasive, and expensive path to take.

Doctor Schoolcraft and his team explain IVF in depth elsewhere in this book so I won't delve into the science and step-by-step procedures here. I will just give you the basic run-through, with a bit of commentary from an IVF veteran who has been through all aspects, the good, the bad, and the ugly.

Basically, the woman receives hormone treatments to increase her production of eggs. (These are the shots that can make your ovaries swell until you feel like a Macy's parade float has been inflated inside you.) Once you have produced more eggs than nature would ever consider, (though some patients may not produce that many) they are retrieved and examined. The mature eggs are fertilized in a lab dish with the male's sperm. The embryos that result are carefully monitored and nurtured and are transferred directly into the woman's uterus with a catheter, usually on Day 5, (or if doing CCS, transferred after results are received).

The patient is put under with anesthesia during this process, which did not thrill this woman child. I have always had a fear of going under and never returning. This was not easy for me to agree to—nor to go through. I was physically and emotionally drained, as are most women.

Bill (Can we call him Mister Wonderful?) was incredibly supportive and patient through all of this. Then again, he wasn't being used as a human pincushion. He gave *me* the twice-daily hormone shots and I will say he was very tender with me when my hormones were raging and I was feeling like a particularly unattractive hot air balloon.

For two months, my ovaries were so swollen I could hardly move without screaming in pain. Still, I was consoled by the fact that my body responded well and fifteen eggs were harvested. Five days later, three fertilized eggs were transferred into my uterus. Three other embryos were frozen in case we needed them later.

❀ *A BREAKTHROUGH AND HEARTBREAK*

A few weeks after the embryo transfer, I finally heard the news I'd long dreamed of getting. Our Chicago fertility specialist called to say that the test showed I was pregnant.

In fact, my hormone levels were so high he thought I might be carrying twins. Bill and I were overjoyed. We immediately began talking about creating a baby room or two. Before we could follow through, however, an ultrasound revealed just one gestational sac.

Still, we were so excited to finally start a family that we made plans to move out of the city and into the more child-friendly suburbs. We wanted a big old house surrounded by trees. Our joyful pursuit of an idyllic family life came to a halt, however, when the next ultrasound visit brought tragic news.

The tech couldn't find a heartbeat and neither could the doctor. I had miscarried.

This was a tough blow to absorb. We grieved as deeply as anyone who has ever lost a child. The only solace was the fact that we now knew I could become pregnant through IVF. We held onto that possibility as we pulled ourselves out of our despair and disappointment.

After taking some time to deal with the loss of that pregnancy, Bill and I decided to try another IVF cycle. With this round of hormone shots, I really earned my stripes for infertility warfare thanks to one of the most dreaded side effects of IVF treatment. My ovaries became overstimulated to the point that they were pushing against my stomach wall, causing incredible pain.

Bill had to take me to the emergency room. The ER doctors gave me painkillers, which I gladly took, but when another doctor recommended a blood transfusion because of my low hemoglobin level, Bill stepped in and said no. He thought it was just too risky, given all we'd read about the dangers of contamination. Fortunately, the doctor agreed to give my body a few hours to catch up and my numbers improved enough to make a transfusion unnecessary.

That scary experience set us back. It took me four months to heal before I could bring myself to try another IVF treatment. This time my Chicago fertility specialist transferred two embryos. The embryos had been above average in quality and everything seemed to go well, but no pregnancy resulted.

When the doctor gave me that bad news, my cries echoed through the beautiful old suburban house we'd bought and restored as our dream home for raising our child. I'd been through so much pain and emotional torment and we had nothing to show for it.

In my anger and grief, I told Bill that I couldn't live there anymore. He didn't fight me on it, even though he'd put so much into renovating the home. At that point, I was ready to give up. I couldn't bear the thought of another failed IVF treatment and all of the shots, raging hormones, and bloated ovaries that came with it.

During this IVF cycle, I'd endured sixty-three shots a month. I could not believe that I had to go through so much to harvest those eggs, and then fail to get pregnant. I'd think: *I'm a good person, and I could give a child the greatest life of all, but yet I can't get pregnant.*

This frustration really hit me when I'd watch reality TV shows about young women who would get pregnant without even trying or even wanting a child. That didn't make sense to me. It seemed like God didn't want me to become a mother, and I thought maybe the time had come to accept that fact and move on.

❀ *A NEW PATH FOUND*

All of my failed efforts in this very private part of my life were public knowledge because they played out on our reality television show. Strangers emailed me, wrote cards and letters, and stopped me on the street or in restaurants to offer sympathy and advice all the time. Most were kind and supportive.

I noticed after a while that several women who'd written to encourage me made recommendations that suggested I try one particular fertility specialist. His name was Doctor William Schoolcraft.

Still, I wasn't ready to return to the battle. In fact, I'd pretty much given up on trying to have a child. But then one day a very attractive and very pregnant woman stopped me as I was getting a mani-pedi in Venice Beach, California.

"I'm sorry to bother you," she said, "but have you ever heard of Doctor Schoolcraft?"

I told her that I'd heard his name from a few other women, but I'd never gone to him. She then launched into a heartfelt story about how she and her husband had tried for years with no luck until she went to this fertility specialist in Denver—and now she was pregnant with twins!

"I'm telling you, you will get pregnant with him. Just try," she said.

Maybe God's plan was to lead me to this doctor whose name kept popping up, I thought. Doctor Schoolcraft must be something special if so many women were recommending him to me.

After the encounter at the salon, I went home and googled him. I found a website with statistics revealing that CCRM has one of the highest success rates for live births in the country. I showed my husband the website and he was intrigued. Bill was delighted that I was looking at another fertility clinic. He thought I'd given up. He asked if we should call and I agreed that it wouldn't hurt. I picked up the phone and began on a new path, wary about getting my hopes up, but excited nonetheless.

❀ *A MILE-HIGH CLINIC*

It had been a long time since I'd actually looked forward to visiting a fertility doctor, but my feeling was that if Doctor Schoolcraft decided he could help us, I was game. I'd already learned that CCRM has a reputation for accepting the most difficult cases. Bill and I were glad for that, because we'd already wasted enough time and gone through enough grief.

I felt refreshed and more hopeful as we drove into the foothills of the Rocky Mountains, just outside Denver where his clinic is located. This was a different sort of medical facility; unlike like any hospital or other clinics, we'd visited. It was very bright and open with a waterfall just inside the main entrance. The setting inspired hope and confidence.

We joined couples from around the country and the world in the reception area. The staff was cordial and very professional. I had the sense that this was a very dedicated and intelligent group of people. No slackers allowed.

Our nurse gave us the rundown of all the tests we'd be taking on this first visit. The list included blood tests, ultrasounds, and full medical workup and exams. The information from these tests would determine the treatment plan that would give me the best possible results.

After months of saying I'd never go through this again, I was surprised to feel anxiety. I guess my desire to be a mother hadn't gone away. Bill, who had quietly hoped I'd change my mind and try again, was enthusiastic and encouraging.

We met late in the day with Doctor Schoolcraft, who was very low-key for a physician with such a big reputation. He was very humble and kind. I'd heard that he was that rare doctor who was also a pioneering scientist in the laboratory. As we talked to him, I felt like his brain was working on several other levels simultaneously.

After reviewing the results of our tests, Doctor Schoolcraft said he thought he could help us. I felt a rush of relief and excitement that surprised me. He was very optimistic about our chances, which was a breath of fresh air for Bill and me.

❀ THE FREEZING OPTION

We hadn't had this much hope in a long time. We even started talking about how nice it would be to have twins. I told Doctor Schoolcraft that I'd need a frequent IVF-er card because if the first one worked at his clinic, I'd be back for more.

By now I was a veteran of the procedure so once I had my meds and my IVF protocol, I was back in the baby-making business. We knew what to expect this time around so the shots didn't bother me and Bill weathered my hormonal storms with his usual patience and kindness.

Doctor Schoolcraft had renewed our hope and once again I found myself thinking about baby bedding and names for boys and girls that worked well with Rancic. After two weeks of hormone stimulation, I returned to Doctor Schoolcraft's clinic and he retrieved my best ever haul of eggs, about eighteen of them. This stage of the game was still a little scary for me since I'd already had a bad experience with hyperstimulation following retrieval.

I expressed my fears to him, and Doctor Schoolcraft's response was unexpected and gratifying. He offered to give me medication to reduce the risk of hyperstimulation, but to do that, he said, my embryos would have to be frozen before transfer.

He explained that if fresh embryos are used they are transferred three to five days after retrieval, but frozen embryos don't have an expiration date. I could take the medication to reduce the chances of hyperstimulation before proceeding. Doctor Schoolcraft assured me that the frozen embryos would work just as well, if not better than the fresh ones.

We chose to put them on ice so I could take the medication to reduce the risk of hyperstimulation. I was glad to go home with the knowledge that I'd eliminated one major concern.

❀ A LIFESAVING TEST

Before we left the clinic, I had one more meeting with a nurse to go over everything before scheduling the embryo transfer. Our show's film crew was with us, but I told them not to worry about filming this boring meeting. The crew said they needed some footage so I let them roll as the nurse read the lists of tests necessary before the transfer.

I'd done my hemoglobin and blood tests, but she realized I hadn't done a mammogram. I tried to wave it off since I had no family history of breast cancer. She wouldn't accept that. A mammogram was required because the hormones I'd be taking could trigger breast cancer in someone who was susceptible, as well as the hormones produced in pregnancy.

I reluctantly agreed to get the mammogram done back in Illinois so I could meet the deadline she'd set. I had no choice if I wanted my transfer done on the date we'd planned, which was just a week away.

My mammogram was done on a Monday and the next day the radiologist called me back. The tech had spotted a speck, maybe because I'd moved slightly. Then the speck showed up on the second mammogram too. The radiologist said it could be nothing, but told me I needed to meet with a cancer specialist. I knew that women often had benign breast cysts and this was just a speck so I wasn't worried, much.

The cancer doctor had me undergo a needle biopsy. When he called me to come in for the follow-up, my husband was out of town, but I had no problem going on by myself. I was that confident.

Still, as I waited in the doctor's office, I began to feel strange. I had the urge to leave, which was unusual for me. I went out into the hallway and tapped the button repeatedly. The doctor's nurse found me waiting for the elevator. I told her that my office had called and I had to go. She insisted that I come back and meet with the doctor immediately.

I tried to convince her, but she gently guided me back into the office, telling me the doctor would be right with me. She said I would be okay. That might have been true in the long term, but in the short term, not so much.

The doctor's first four words were all I needed to hear before feeling sick: "Giuliana, I'm sorry but...." I felt like someone waiting for a punch to land, I braced myself for the news that I already knew would change my life dramatically.

"You have breast cancer."

My first thought was that I was going to die, soon. I didn't know then that if your breast cancer is detected early enough there is a 98 percent survival rate. The doctor tried to reassure me, but I had gone into survival mode. I decided that rather than deal with this new setback I had to go to work.

I gutted it out that day, hosting *E! News*. I don't know how I pulled it off. Afterward, I ran into my dressing room and cried hysterically. Bill came to the rescue. Later that night, he picked me up off the floor and said, "Hey, we're making a plan here."

❋ *A NEW FIGHT*

I let my husband take over because I was lost in another dimension of despair and grief and disbelief. I wrote in-depth about all of the drama that ensued in my own most recent book, *Going Off Script: How I Survived A Crazy Childhood, Cancer and Clooney's 32 On-Screen Rejections*. The important thing to note for this book is that I survived. It's important to note that I might not have survived at all if Doctor Schoolcraft and his team had not insisted that I get a mammogram. Because of their insistence, my cancer was detected early in its development.

There was no history of breast cancer in my immediate family, so I would have waited until I was forty years old to get a mammogram. That was what I'd been told to do many times. Instead, I went in for a mammogram kicking and screaming at thirty-six because I wanted to have a baby and it was one of the requirements of Doctor Schoolcraft.

When I went on the *Today* show with Ann Curry to encourage other women to be more vigilant, Ann made the point that 85 percent of those diagnosed with breast cancer have no prior history. I was learning more than I'd ever known about cancer, and my education had only just begun.

Shortly after I first went public with my cancer on that show, my doctor called with even worse news. Typically, doctors check both breasts as a precaution. My physician found something suspicious in the other breast, a small spot, and the pathology report said it too was cancerous.

With that discovery, I regretfully became a rarity—a low-risk candidate for breast cancer who had it in both breasts. I'm told this occurs in less than five percent of women.

I underwent a double mastectomy on December 13, 2011. I've written and talked about this in other places so I won't go into it here other than to say the breast cancer definitely set me back in my long and difficult journey toward parenthood.

❀ STAYING STRONG

Women who have difficulty conceiving often talk about feeling trapped in this endless loop of crises, disappointments, and grief. They often feel like life is constantly knocking them down and daring them to get back up.

You have to be a very strong person to keep hope alive. Amazingly, most of the women I know find that strength, but few do it on their own. You should welcome all the support you can get. Tap into whatever sources are available to you. My husband's mantra through all of our trials was, "You can have fear, or you can have faith."

We chose faith, and we needed it in abundance. Gratitude is another wonderful treatment for grief, anger, and fear. And, as strange as it may sound, I came to thank my infertility for saving my life. Because of it, I went to the Colorado Center for Reproductive Medicine and the team there would not let me slide on taking the required mammogram.

I was horrified at the results, of course. The surgery terrified me and the recovery was a bitch, but I survived to continue my quest for motherhood. I vowed not to give up; and how could I? This baby I was seeking had saved my life.

Focusing on the positive helped me retain my sanity. Whenever I was so sad, when I was on the floor kicking and screaming and crying, the best way I could pull out of it was to remind myself that even with this happening, I am alive and still blessed in many ways.

I found gratitude by thinking: *Yes, I have breast cancer. Yes, I had to have a double mastectomy. But, you know what? Thank God I'm going to a good doctor. I'm so fortunate I found it early. I'm so fortunate to have Bill.*

❋ *A NEW PLAN*

As an eternal optimist, okay, maybe as a diehard Pollyanna, I'd always said that God had a master plan that, for some reason, included my failed IVFs and a miscarriage. I still believe that because if I'd gotten pregnant with undiagnosed breast cancer, I could have become a lot sicker.

Still, cancer did not make my life any easier. After my surgery and the biopsies of the tumors, my doctors realized this was a particularly aggressive type of cancer. They suggested I undergo chemotherapy.

I didn't want to do it. I had fought so long to have my own child. I was afraid the chemo would wipe out any hope. After a great deal of discussion and research—Bill put together a very distinguished group of consultants to help me weigh my options. My doctors recommended I take the anti-estrogen drug tamoxifen for five years to reduce the risk of a cancer recurrence.

That meant I'd be forty-two by the time I was off the anti-cancer drugs. By then, I'd have to worry whether there would be any viable eggs to be harvested for IVF. So that was another big concern. Before I had time to let the implications sink in, Doctor Schoolcraft stepped up with a new plan built around an option that I hadn't considered: "gestational surrogacy."

My initial thoughts, based on my own misunderstanding, were not enthusiastic. I imagined surrogacy involved a strange woman having a child with my husband, a child that was more hers than mine.

Okay, maybe I overreacted. Bill and I had discussed surrogacy years earlier and dismissed the idea because it was a different sort of surrogacy. I'd known someone whose seemingly benevolent surrogate had turned out to be the mother of all shakedown artists. We didn't trust a process that gave a stranger so much power over our lives.

❀ A CO-OP FOR CONCEPTION

Doctor Schoolcraft swiftly put to rest most of our fears and concerns. He explained that *gestational* surrogacy meant that Bill and I would make a baby—our baby—and then have another woman carry the child in her womb. The baby would be ours genetically, biologically, and legally.

This was a different kind of deal. In effect, we'd just sublet a cozy baby room from some extremely generous stranger. As our doctor noted, we had frozen embryos in storage, so this was a path we could quickly move along.

The truth was that my cancer and the follow-up treatments had eliminated most other options. Bill had some stronger doubts than me, but we decided to interview a few surrogates just to see how we felt about them.

Doctor Schoolcraft's team recommended a "matchmaker" agency that paired couples with surrogates. We were surprised when they called us just a few days after our initial meeting. They already had found a surrogate who'd been vetted and available because another couple she'd been working with had somehow become pregnant on their own.

This appeared to be God's sign for us to move forward. The poor surrogate didn't know what she was in for. By the time we met, I had filled a yellow legal pad with questions for her, everything from her eating habits to medical history to her preference of the Cubs versus the White Sox—well, maybe not that last one, though I did consider it.

Delphine turned out to be a pretty and petite thirty-three-year-old native of France with a great job, a husband, and two little boys of her own. She was educated and intelligent and thought surrogacy was a wonderful gift she could give another mother. She had never heard of our reality show, which I secretly liked. We bonded and within a short time, signed a surrogacy agreement.

❀ *THE ULTIMATE GIFT*

While I was recovering from my mastectomy, Delphine had the embryo transfer with our embryo at Doctor Schoolcraft's clinic. Bill and I were hosting a Times Square gala on New Year's Eve when I received another phone call from Doctor Schoolcraft that would prove to be life changing. While others waited for the giant ball to drop and mark the New Year, my husband and I were waiting to learn whether the embryo transfer had been successful. I needed some good news and the good doctor came through.

"Congratulations, you're having a baby," Doctor Schoolcraft told us.

We brought in the New Year 2012 with a kiss broadcast live around the world and Bill declared that this was our year. A year of much better news and events, I hoped.

I was fighting cancer and fighting for a family too. We focused on having a positive experience with our surrogate. Bill and I were with Delphine in Denver for every doctor's visit. The sonogram introduced us to our son and I was overjoyed.

We grew close to Delphine and her family. I respected her privacy, and she was very good about texting me to keep me posted on the baby's activity. I was sad that I couldn't carry my own child and worried about him being carried by another woman. It was easier if I thought of him as a three-year-old towhead with floppy hair, chubby cheeks and dimples. I looked ahead and focused on images of Bill and I playing with our toddler to keep myself calm during those nine months.

As the baby's due date approached in the last week of August 2012, Bill and I flew to Colorado for the delivery. We went hiking in the mountains first to work off our nervous energy, but we still acted goofy in the waiting room until a nurse told us the baby was about to arrive. We were in the delivery room, standing at Delphine's head, offering encouragement through our tears while her husband and a nurse coached her.

Our son came out in full roar, announcing his hard-won arrival. Bill cut the cord and before I knew it, I was holding our child—our ultimate gift—against my chest, skin to skin, hearts beating together.

Edward Duke Rancic entered the world on August 29, 2012. I had a child to love, finally, and I wanted to spend as much time as I could with him. The fight with cancer had made me more aware of the preciousness of life and how little time we have to love and be loved.

I have continued to work in television and other endeavors, but I've greatly reduced the time I am away from my son and husband. Bill and I also have been building businesses so that I don't have to travel so much. Work smarter, not harder, is my new motto.

❋ STRIVING FOR SIBLINGS

After Duke's arrival, we still wanted to provide him with siblings. Three perfect embryos remained frozen after his birth. I even harbored fantasies of being a full-time mom who could take her kids to school, hand in hand, go to their soccer games and read to them while cuddling on the couch.

In 2014, we asked Delphine if she would help us have another child. She agreed and Doctor Schoolcraft transferred two of the three frozen embryos into her uterus. We were thrilled when he called to say that Delphine was pregnant. Yet, our joy was short-lived. Eight weeks later, Doctor Schoolcraft told us that the embryo had detached from the uterine sac and the pregnancy had ended.

Doctor Schoolcraft was obviously disappointed and sad too. He told us that it just wasn't logical, but random chance is part of life. "Ultimately," he said, "it's up to the universe." He added a hopeful note, however, saying, "It doesn't mean the next time won't be a success."

My life has been blessed, yet like most people, I've known grief and great depths of sadness. I've learned to let the tears and sorrow run their course, and I've learned to surrender to whatever bigger plan is playing out. I no longer try to understand why. Instead, I wait for the clarity to come to me one day.

Once again we dealt with this deep disappointment and once again, we emerged from the grief and decided to keep fighting. This time, however, we would be using our last frozen embryo.

We felt it was worth it. Our little Duke was growing so fast, and we want him to have at least one sibling. Delphine agreed, and Doctor Schoolcraft transferred our last embryo.

She became pregnant, but on New Year's Eve 2014—three years after Bill and I had learned on the same day and hour that we were going to have baby Duke—Doctor Schoolcraft called again, but with very sad news.

"I'm sorry," he said. "There is no baby."

Our "last chance" pregnancy with our own embryo ended six weeks into the pregnancy. You'd think I would be numb to the pain by that point, but it was worse than ever. We thought this time would definitely work. It was our last shot and so it was the toughest blow, by far, of all we have experienced in this journey.

Bill, always my source of strength, consoled me, saying that we already have a complete family. Still, I felt a great loss. Those feelings may never fade altogether. They have been allayed somewhat by a greater awareness and gratitude for the child we do have.

I realize more than ever just how lucky we were to bring him into the world. Once again, I am reminded that as painful as they may be, the disappointments and dark times can give us a greater appreciation for the blessings and light in our lives.

Choosing to focus on that light, I keep my heart open to all possibilities. There are many brilliant women and men doing wonderful things in the field of reproductive science. They give me hope. They give hope to many others, as well.

I've written before that life is never the perfect fairy tale, but sometimes we forget that life itself is a gift. All of the ups and downs we go through, the laughter and the tears, are part of the journey. The suffering we endure can give us a greater appreciation for the blessings we receive. I've learned to be grateful for every moment I'm allowed to share with my husband and our son.

Doctor Schoolcraft and his wonderful team probably saved my life. I know for sure that in helping us to have a son, they made my time here all the richer and more meaningful. If you are still trying to overcome infertility, I hope my story has helped inspire you and given you strength. My heart and my prayers go out to all women engaged in this fight. Know that you are stronger than you may think, but please don't isolate yourself in the hard times; welcome all of the love and support you can find.

CHAPTER TWO

IN VITRO MATURATION, A BETTER OPTION

BY DOCTOR WILLIAM SCHOOLCRAFT

Shannon came to our clinic after having difficult a experience at another facility. After undergoing a traditional in vitro fertilization procedure there, she had to be hospitalized in an intensive care unit. You see, Shannon has polycystic ovarian syndrome (PCOS), which affects as many as five million women in the United States. Considered one of the most common causes of infertility, PCOS is characterized by high androgen levels, small cysts in the ovaries, and irregular periods.

The hormonal environment in this condition doesn't allow the eggs to fully mature and ovulate regularly. Their follicles may start to grow and build up fluid, but ovulation doesn't occur. Some of the follicles may remain and become cysts. Ovulation does not occur and the hormone progesterone is not made. Without that progesterone, the menstrual cycle is irregular or nonexistent.

When women with PCOS are subjected to controlled ovarian hyperstimulation, which is the usual procedure with IVF, their bodies overreact, putting them at risk for ovarian hyperstimulation syndrome. There are so many follicles available to grow in response to the medications that enlargement of the ovaries can become extreme. This can be a serious and even life-threatening complication.

That is what happened with Shannon. She had over-stimulated after a trigger shot of HCG, the hormone used in IVF to stimulate the release of eggs. She developed huge ovaries and a large amount of free fluid in her abdomen. This free fluid then crept into her lungs, causing respiratory compromise, and it increased the thickness of her blood, leading to a blood clot.

Needless to say, after a long recovery she was terrified of attempting IVF again. We discussed her high risk with drug therapy, and I explained that there was an alternative to IVF that used no medications whatsoever. She was relieved to hear about this more recently developed option, which is called "in vitro maturation," or IVM.

This more cost effective option requires less medication because immature eggs are first collected from the ovary, then matured in a laboratory. Next, the eggs are fertilized, and the embryos that form are matured in the lab before being transferred to the mother's womb. Although IVM is a research protocol and most clinics don't have the resources to perform it, patients who are at risk for hyperstimulation with the traditional IVF treatments may find it a good alternative.

❧ *A SAFER ALTERNATIVE FOR SOME PATIENTS*

In vitro maturation (IVM) was developed to provide a safer alternative to conventional in vitro fertilization. The standard in vitro fertilization process is hampered by the requirement for controlled ovarian hyperstimulation of the patient's ovaries. This requires them to be on medication sometimes for twenty days or more, taking several injections per day and going in for frequent ultrasounds and blood monitoring. This traditional IVF treatment is a painful and tiring experience preceding the egg retrieval.

Doctors and patients alike have long wished all of these IVF steps could be skipped entirely, which is indeed the case with the less demanding in vitro maturation treatment. In this procedure, the woman's eggs (oocytes) are collected from small follicles, which are the little sacs of fluid that contain immature eggs. Typically, around day nine to day ten of the menstrual cycle, the eggs are aspirated just as is done with standard IVF.

IVM typically allows more than half of the harvested eggs to become mature. Then, we can fertilize the eggs to create viable embryos. While pregnancy rates are not quite as high as with conventional ovarian hyperstimulation, or standard IVF, they can exceed 40 percent per cycle in the best infertility treatment centers.

❀ CANDIDATES FOR IVM

Younger women under thirty-nine with a good number of follicles and young women who have received a cancer diagnoses with the need to preserve their fertility prior to cancer treatments, are among the candidates for IVM. But there are also two other types of patients who could benefit significantly from this alternative to IVF.

The first group includes patients like Shannon with polycystic ovarian syndrome. Obviously, those PCOS patients who want to have children are grateful for a procedure that eliminates drugs that cause them to develop ovarian hyperstimulation. PCOS is the most common type of patient treated with IVM. Because these patients have so many follicles, many eggs can be harvested which is a real benefit for IVM.

Caroline represents the other type of patient that might benefit from in vitro maturation. These women have the opposite extreme in ovarian function compared to the PCOS patients. She is considered a "poor responder," which is the rather harsh sounding medical designation for women who respond poorly to the standard controlled ovarian hyperstimulation drug regime in IVF.

In the past, many of these patients have received four thousand dollars to five thousand dollars in medication over ten to twenty days, only to produce one or two eggs more than they would have in a natural cycle. This is not a very cost effective way to help women with fertility challenges produce more eggs.

There are also concerns that such high doses of medication might adversely affect the function of those few eggs that are harvested. That is why IVM has been considered as an alternative for Caroline and other women who don't respond well to standard IVF.

To retrieve the eggs from the follicles, we do a follicular aspiration procedure in which a needle is introduced through the vagina and guided inside the follicles using ultrasound. Anesthesia is used for the patient's comfort. In such cases, the lead follicle can be aspirated, again at twelve to fourteen millimeters, often yielding a mature egg. The smaller resting follicles can also be aspirated, yielding immature eggs that can then be matured in the lab.

With this method, patients can undergo several aspirations because the procedure is less complicated, not calling for the use the complicated high dose drug protocol. The outcome may be as good or better, and certainly is more cost effective. Furthermore, it is far easier on the patients' mentally, emotionally, and physically, as well as being less time consuming.

Caroline was in her early forties when she came to us for help. She had been through two conventional IVF attempts with high doses of medication. In both cycles she was thought to have three to five follicles, but after egg retrieval she was deeply disappointed to learn that no eggs were harvested. This happened to her not once, but with two different cycles, requiring an expenditure of over ten thousand dollars each time, not to mention a month of injections and tests that produced no eggs.

When Caroline came to us for treatment, I told her that controlled ovarian hyperstimulation and the standard drugs for IVF were clearly not going to work for her. In the past, the next logical option would be to look for an egg donor, but Caroline still wanted a child with her genes. I explained to her that in vitro maturation might be an option worth exploring.

After I explained the process, Caroline made the decision to try IVM. She began with an egg retrieval around cycle day nine. We obtained four eggs. One of them came from a lead follicle twelve to fourteen millimeters, and it was mature at the time of retrieval. Three more retrieved eggs matured within two days in our lab.

Amazingly, all four eggs fertilized and grew to the blastocyst stage. After comprehensive chromosomal screening, it turned out that one was a euploid or a chromosomally normal blastocyst. Six to eight weeks later, Caroline underwent a frozen embryo transfer and conceived successfully.

This was quite dramatic for several reasons. First of all, she had diminished ovarian reserve, so the fact that she was able to achieve a pregnancy with IVM was remarkable. On top of that, she'd never had an egg even retrieved with standard IVF. Needless to say, we were all very happy for Caroline, and she was thrilled.

❁ *THE LAB WORK*

From a laboratory perspective, IVM is really a science unto itself. There are only a few centers around the world equipped to provide the cutting edge technology, so the lab is really the main driver for in vitro maturation technology. While standard culture conditions can take mature eggs and allow for fertilization, embryo cleavage, and growth all the way to the blastocyst stage, those conditions are not optimal for the maturation of oocytes. Our foundation's researcher, Doctor Rebecca Krisher, along with other scientists around the world, are working to optimize the culture conditions, growth factors, and other requirements needed by immature eggs to mature successfully in vitro.

The lab team has to allow the cytoplasm or the liquid in the egg to undergo adequate maturation. The same is true with the nucleus, which must mature to what is known as a metaphase II state, at which time the egg is ready for fertilization. Special nutrients in the culture media have been shown to optimize IVM, as well as hormones and other growth factors.

Another interesting aspect of in vitro maturation is that it obtains the egg from the ovary before it undergoes its meiotic divisions, the process by which reproductive cells (gametes) are formed. It is during the first and second meiotic divisions of the egg that most chromosomal errors or mix-ups occur. This is a particular problem in older women.

The goal of researchers is to identify and provide optimal conditions in the laboratory for the first and second meiotic divisions to take place. This might allow for these steps to go in a more orderly fashion, leading to more genetically normal (euploid) eggs. This of course would be a very big bonus for older women because most of their harvested eggs are chromosomally abnormal.

Once the eggs are matured in the laboratory, which usually takes one to three days, fertilization is accomplished through ICSI, the standard IVF laboratory procedure in which a single sperm is injected directly into a single egg to achieve fertilization. Embryo culture is then undertaken, either for three or five days, at which time embryos can undergo genetic screening and be frozen.

Embryo freezing is typical with IVM cycles because the patient has not ovulated before retrieval. Therefore, the patient's uterus is not prepared for a fresh embryo transfer. We will outline the advantages of frozen embryo transfers in the following chapter. For now, just be aware that they have certain advantages compared to fresh cycles and therefore offer a nice complement to the in vitro maturation procedure.

Once embryos are created and frozen, the patients go on a controlled estrogen and progesterone cycle to prepare the uterus for embryo transfer. While IVM is a new procedure and there is limited data on birth outcomes, the babies reported thus far from in vitro maturation do not appear to have an increased risk for congenital anomalies or any other types of defects compared to standard in vitro fertilization conceptions.

❀ *AN OPTION WORTH CONSIDERING*

The two patients I've mentioned in this chapter, Shannon and Caroline, have fertility challenges at opposite ends of the spectrum. Shannon has polycystic ovaries and produces an excess number of eggs, while Caroline has very few eggs. Both women experienced great difficulty and little success with the standard IVF, but both benefited greatly from IVM.

This method has been studied in reproductive biology for many decades, particularly in animal models. However, it has been relatively recent that in vitro maturation has gained a degree of success and efficiency to make it a consideration for patients seeking infertility treatment. In the coming decade, I would predict that better culture conditions and a better understanding of what the requirements are to mature an egg successfully outside the body will lead to greater use of this technology. Indeed, some scientists have even suggested it may replace conventional IVF treatment.

As the science of in vitro maturation continues to evolve and our culture conditions improve, I believe there may come a day where we routinely apply IVM to all patients and eliminate the need for expensive, painful, and risky fertility drug treatment. That would truly be an advance for our field and our patients.

CHAPTER THREE

FROZEN EMBRYOS ARE EVEN BETTER THAN FRESH

BY DOCTOR JAMES L. NODLER, M.D.

Deanna always knew she wanted to be a physician. She loved science and helping people, so applying to medical school after college was a no-brainer. She worked hard, and made it to the top of her class, earning the highest grades on difficult clinical tests and board exams.

When the time came to choose a specialty, she applied to dermatology, which is an ultra-competitive area. Even so, Deanna made it into her chosen specialty. Her career path as a dermatologist was off to a great start. She'd been out of school two years when she met the man she would marry. After settling down as a married couple, they were ready to start their family.

With her medical training and good health, Deanna thought she would have no problem getting pregnant at the relatively young age of thirty-two. This is a common assumption among young women, one that we try to discourage. Instead, we recommend that women who are twenty-five or older undergo a fertility assessment exam so that they know if they might face any challenges becoming pregnant one day.

Deanna had taken birth control pills for more than ten years during her medical training because she didn't want an unplanned pregnancy derailing her progress along her career path. When she stopped the pills in an effort to become pregnant, she noticed that months passed without a period. She tried to use home ovulation predictor kits, which measure the rise of luteinizing hormone before ovulation, and she could never get a positive result. She knew this meant she was likely not ovulating, or releasing an egg.

Concerned, Deanna made an appointment with her good friend, who was an obstetrician and gynecologist. The OB-GYN diagnosed her with PCOS—polycystic ovarian syndrome—a common disorder that can make it difficult to have regular periods and to get pregnant.

Blood testing also showed that she had elevated levels of male hormones such as testosterone, and an ultrasound of her ovaries showed they contained dozens of small cysts. Her doctor reminded her that women with PCOS can usually get pregnant, but often need medical help to release an egg that can be fertilized.

Following her doctor's advice, Deanna underwent six months of treatment with an oral medication called letrozole that can help women ovulate. Still, she did not become pregnant.

Deanna was growing desperate with each passing month. Her anxiety over her infertility was affecting her work as a dermatologist as well. She was beginning to lose hope when she was referred to our clinic. We discussed the treatment options with her and she decided that in vitro fertilization would give her the greatest chance of success.

Deanna was thrilled to hear that the predicted live birth rate with IVF at her age was nearly 60 percent per attempt. We told her she would respond well to the treatments after seeing the results of a blood test called AMH, or anti-mullerian hormone, and with ultrasound measurement of resting follicles or cysts in her ovaries.

The process of IVF usually involves the patient taking fertility hormone injections to induce multiple eggs to grow at the same time, which improves the odds of success. After approximately ten days of stimulation the eggs are ready for harvest or retrieval by a procedure known as transvaginal aspiration.

The eggs are fertilized and the resultant embryos can be transferred with either a fresh transfer, or frozen to transfer at a later time. Deanna had some common concerns. She had fears of ending up with too many fertilized eggs and becoming an "Octo-mom." We hear this a lot. The truth is we can control multiple pregnancy rates with IVF by transferring only one embryo in younger women such as herself.

However, we must take special care with women such as Deanna who are "high-responders." We don't want them to become over-stimulated in the process of IVF. High responders often have much higher levels of estrogen, progesterone, and other hormones than the body normally experiences. Women like Deanna can become over-stimulated and develop a condition called ovarian hyper-stimulation syndrome, or OHSS.

This occurs when high levels of estrogen and another protein called VEGF (vascular endothelial growth factor) cause blood vessels in the body to be "leaky" leading to fluid buildup in the abdomen and electrolyte imbalances. These high hormone levels seen after a "fresh" IVF stimulation have been shown to cause several other problems such as lower embryo implantation, lower pregnancy rates, higher ectopic pregnancy rates, and problematic fetal outcomes including lower birth weight and more frequent preterm delivery.

We are fortunate that stimulation protocols and lab techniques have improved to the point where we can offer patients a high chance for a pregnancy as well as minimized risk for the complications associated with OHSS.

❀ OHSS AVOIDANCE

Freezing Deanna's embryos sharply decreased her odds of developing ovarian hyper-stimulation syndrome (OHSS). This can be a severe complication of IVF associated with a spectrum of symptoms ranging from mild bloating to severe electrolyte imbalances and blood clotting. It is not uncommon for patients with OHSS to have to undergo a paracentesis, in which a needle is used to drain fluid build-up in the abdomen.

Women with OHSS can be extremely uncomfortable and short of breath during a time where they only want to celebrate a long-awaited pregnancy. The exact cause of OHSS is unknown, but it is seen most commonly in women like Deanna who produce many eggs and have high estradiol levels.

The combination of a "Lupron" trigger and avoidance of a fresh embryo transfer are the keys to minimizing OHSS. The following month, a frozen embryo transfer can be completed when the ovaries have returned to normal size and the risk of OHSS is eliminated.

❀ A NEW OPTION WITH MANY BENEFITS

We often quote the motto of our clinic, which is, "The only measure of success is the birth of a healthy baby." With patients like Deanna, we can help to accomplish these goals by using a technology called "embryo freezing," also known as "cryopreservation."

Embryo freezing has been around since the early 1980s, debuting not long after the advent of IVF. Early IVF patients would often produce multiple embryos, but not all of them could be transferred at once due to fears of multiple pregnancies. For this reason, excess embryos had to be discarded, and if the original "fresh" IVF stimulation was not successful, patients would have to start the process again from scratch.

Embryo freezing originally involved letting the embryos grow for three days after fertilization then slowly cooling them before immersing them in liquid nitrogen. This process led to the first live birth in Australia, from a frozen embryo in 1984. While effective, "slow freezing" led to pregnancy success rates that were significantly lower than fresh embryo transfers because the freezing process caused ice crystal formation that could damage the embryos. A new process called vitrification was developed. In this method, embryos are typically grown to the blastocyst stage, which occurs five days after fertilization. Then they are cooled in an ultra-rapid process before submersion in liquid nitrogen. This ultra-rapid freezing process allows embryos to be frozen in a glass-like state, avoiding ice crystal formation and preserving embryo quality.

Embryo freezing offers multiple advantages to the process of IVF. As previously mentioned, embryo freezing allows excess embryos from a fresh IVF cycle to be stored and used in the future.

Once Deanna decided to proceed with IVF, we suggested that she *not* have a fresh embryo transfer. Our recommendation was for her to freeze all of her embryos after growing them for five days after fertilization so they reached the more mature blastocyst stage.

Several factors led us to this recommended course of treatment. As noted previously, Deanna is a "high responder," making many eggs in IVF stimulation. She was fortunate to have twenty-five eggs retrieved. Twelve grew to the day five, blastocyst stage and were frozen for future use.

❀ STOPPING THE CLOCK

Fertility declines in women as they age for two main reasons— they have fewer eggs remaining and the quality of eggs diminishes, often due to the accumulation of chromosomal or genetic errors that increase over time. Frozen embryos do not continue to age or accumulate chromosomal defects while they are frozen.

When Deanna froze her embryos at age thirty-three, she was comforted to know that they would remain that age indefinitely. Her pregnancy success rate with these embryos will remain similar to that of a thirty-three-year-old woman even when Deanna reaches her forties. Women can use their frozen embryos to achieve pregnancy if their first attempt at IVF is not successful, or if they would like a sibling after successful IVF several years down the line.

A frozen embryo transfer cycle is much easier and cheaper than a fresh IVF stimulation. Patients in a frozen cycle pay significantly less for fertility medications, and do not have to undergo the egg harvesting procedure.

❀ RISK REDUCTION

In Deanna's case, our plan for freezing all of her embryos and not performing a fresh transfer minimized her risk of serious health problems from OHSS. Women who undergo the process of IVF do so in order to increase their chances of pregnancy, obviously. Recent research studies indicate that embryo freezing may increase implantation rates and pregnancy success when compared to fresh embryo transfer, especially in patients such as Deanna with high estradiol and progesterone levels.

There are many explanations for why pregnancy success may be lower in a patient following fresh embryo transfer directly after IVF stimulation. Reasons generally revolve around high estradiol and progesterone levels creating an inhospitable environment for the developing embryo.

Studies have shown high hormone levels to be associated with structural changes in how the endometrium, or uterine lining develops. It is important to remember that the goal of a fresh IVF stimulation is to get as many eggs as is safely possible, and not necessarily to help the endometrial lining, to form in a physiologic, or natural way.

The endometrial lining is like a bed or cushion for the developing embryo and its correct formation is vitally important to pregnancy success and subsequent fetal growth. At the time of implantation, the endometrial lining must be thick and active for a pregnancy to grow. A complex network of uterine blood vessels must form to feed the developing pregnancy.

The uterine environment after a fresh IVF stimulation can cause abnormal growth and activity of the endometrial lining, which can lower pregnancy rates. Studies have shown high hormone levels to affect uterine blood vessel formation as well.

With a frozen embryo transfer cycle, all attention is focused on growth and development of the endometrial lining. Hormone levels are closer to physiologic, or natural levels. This allows for more natural growth of the endometrial lining, which can lead to better implantation and pregnancy rates.

Another emerging area of research focuses on how the developing embryo synchronizes with the endometrial lining. This topic is often termed "endometrial receptivity." The lining physically changes depending on the time in the menstrual cycle, and the concentration of hormones that are stimulating it.

At the beginning of a cycle, immediately following menses, the endometrial lining is very thin and not hormonally active. After ovulation occurs mid-cycle, the lining grows thicker, secreting hormones and other factors. The lining must be at an appropriate "phase" of development in order for implantation and pregnancy to occur.

In a patient like Deanna, estradiol levels following fresh IVF stimulation can be elevated, which subsequently leads to elevated progesterone levels. These elevated progesterone levels can advance the "phase" of the endometrial lining, which can make the lining less receptive to the developing embryo and lower pregnancy rates.

The advantage of embryo freezing is that normal hormonal concentrations promote improved endometrial receptivity, which can improve pregnancy rates.

● *LOWER RATES OF ECTOPIC PREGNANCY*

Embryo freezing can also be beneficial for the health of women undergoing IVF. Multiple studies have demonstrated a lower rate of ectopic pregnancy in women who had frozen embryos transferred than in women who had fresh embryos transferred directly after an IVF cycle.

An ectopic pregnancy is a pregnancy that implants anywhere outside of the uterine cavity. Ectopic pregnancies are more likely to occur following spontaneous pregnancy than with IVF. The most common location of an ectopic pregnancy is in the fallopian tubes because they have a very narrow diameter. A developing embryo may get "stuck" in a tube, especially if there has been previous tubal damage.

Ectopic pregnancy is a very serious condition for several reasons. First, an ectopic pregnancy that develops in a fallopian tube can break open the tube and rupture causing massive uncontrolled bleeding inside the abdomen. A developing pregnancy grows much larger than a fallopian tube, and when it does, rupture can occur.

Early pregnancies following IVF are typically followed very closely, reducing the risk of serious complications in a woman with an ectopic pregnancy. However, uncontrolled bleeding with an ectopic pregnancy can cause serious injury or death. Ectopic pregnancy can also lead to significant delays in fertility treatment, which can be devastating, especially in older women.

❀ THE SINGULAR ADVANTAGE

Freezing embryos allows us to promote single embryo transfer, which greatly improves the odds of having a singleton pregnancy, rather than a twin or triplet pregnancy. While advances have been made in the delivery and care of twin pregnancies, it is well known that singleton pregnancies are healthier for both moms and babies.

Multiple pregnancies have a higher risk of preterm delivery and low birth weight infants. These premature infants can be at increased risk of several conditions such as cerebral palsy, hearing and vision loss, and mental retardation. Women with multiple pregnancies are at increased risk of preeclampsia, resulting in dangerous elevations to blood pressure during pregnancy, as well as gestational diabetes during pregnancy.

The development of reliable and successful embryo freezing has allowed fertility specialists to transfer fewer embryos, often just one in women less than thirty-five years old. The goal is for a healthy pregnancy and delivery.

❁ *BETTER SCREENING WITH EMBRYO FREEZING*

Embryo freezing also promotes healthy pregnancies because it allows us to perform comprehensive chromosome screening (CCS). CCS is made possible by embryo freezing because it allows physicians time to test the embryos and prepare the uterine lining for a frozen transfer.

We will discuss CCS in detail in the next chapter. Briefly, when chromosomally normal embryos are later transferred in a frozen transfer cycle, pregnancy success rates are higher and miscarriage rates are much lower. Also, CCS allows fertility specialists to transfer fewer embryos, which again, promotes lower risk of multiple pregnancy.

❁ *FROZEN EMBRYOS ARE SAFER*

Naturally, one would think that the process of freezing could damage embryos. However, the intense care taken with embryos through the process of freezing, along with new technologies like vitrification, or ultra-rapid freezing, help to prevent damage to embryos and resultant babies. Studies surrounding embryo freezing were first focused on whether freezing embryos is safe at all for embryos and developing babies.

Fortunately, many studies have shown that the process of embryo freezing does not lead to higher rates of abnormalities such as preterm delivery or low birth weight in babies born after assisted reproduction.

Interestingly, multiple studies have shown that frozen embryo transfers are associated with a higher likelihood of better fetal outcomes when compared to fresh embryo transfers. This improvement in fetal outcomes following frozen embryo transfers has been seen even in the absence of CCS, or chromosomal screening as discussed above. The reason for the improvement in pregnancy and fetal outcomes with frozen embryos in comparison to fresh embryos likely centers around the markedly increased levels of estrogen, progesterone, and other co-factors seen with fresh IVF transfers and the associated ovarian stimulation. A frozen embryo transfer cycle is designed to much more closely mimic the hormone levels seen in natural pregnancy.

The specific fetal outcomes that may be improved following a frozen embryo transfer when compared to a fresh transfer are a decreased rate of preterm delivery, lower incidence of preeclampsia, and higher fetal birth weight. These outcomes are some of the most important in determining the health of the baby that couples with infertility have worked so hard to have. The higher hormonal levels seen after fresh IVF may also be associated with metabolic diseases such as diabetes that could be developed later in life.

❀ THE FAMILY PLAN

We are happy to report that embryo freezing has already been very helpful to many patients, such as Deanna, in achieving their ultimate goal of healthy babies. Deanna had one frozen embryo transferred, and did not have any signs or symptoms of ovarian hyperstimulation even though she made many eggs and had a high estradiol level in her fresh stimulation.

She was fortunate to become pregnant on her first attempt. She gave birth to a healthy son, Jack. She has other high quality embryos frozen, which she plans to use over the next several years to keep growing her family.

Without the development of embryo freezing, a patient like Deanna may have felt pressed to attempt pregnancy again quickly to complete her family before her egg quality started to decline with age. She feels more confident now that she can build her family on her own terms.

CHAPTER FOUR

THE BENEFITS OF GENETIC TESTING

BY MANDY KATZ-JAFFE, PHD, DIRECTOR OF GENETICS AT CCRM

Lisa has said that she and her husband, Jack, "were twice blessed" by a type of genetic testing offered by our clinic.

"Without this technique, it could have been just Jack and me looking at pictures on the wall instead of raising a family," she said.

The process Lisa referred to is Preimplantation Genetic Diagnosis (PGD), an extremely useful and relatively new tool for those seeking to have a child. Do not allow the dry scientific name fool you. PGD is an incredible scientific advancement that saves lives and spares families great anguish. Our clinic performed one of the world's first PGD trials.

This wonderful, yet controversial tool played a pivotal role in what is probably the most compelling and dramatic case we've ever had—one that became a major news story around the world: Lisa and Jack Nash used genetic diagnosis to have two children—and to save one of them.

Despite the complexity of PGD, it has become a hot topic in our field, and one of the internet's most searched and discussed items related to fertility. The science of PGD begins with this simple fact: Birth defects occur in nearly one in twenty pregnancies. These defects range from minor physical abnormalities to serious genetic disorders and mental retardation.

Some couples, including Coloradans Lisa and Jack Nash, have a greater than average risk for creating a child with a serious birth defect. Our clinic helped pioneer the use of genetic diagnosis, in combination with in vitro fertilization, to reduce the risk for those couples. PGD allows for the detection of disease-causing mutations and chromosomal abnormalities in embryos prior to conception. We have found it to be a vital tool for a wide range of our patients, especially for older women seeking to become pregnant, and for those with known chromosomal abnormalities, gene disorders, and a history of miscarriages.

We knew PGD was an important tool, but the true significance became apparent when we met a young lady named Molly, daughter of Lisa and Jack. She was born on July 4, 1994 in Denver, Colorado. Molly was missing both thumbs. Her right arm was 30 percent shorter than her left. These were recognized as symptoms of Fanconi anemia, a blood disorder characterized by a deficiency of red blood cells, white blood cells, and platelets.

Those born with Fanconi anemia have increased risk for cancer and congenital birth defects. The disease is often associated with cardiac, kidney, and limb abnormalities. Short stature is common in children and adults with Fanconi anemia. They often bleed and bruise easily and suffer from hormonal and fertility problems. Many with this disorder do not survive beyond young adulthood as a result of leukemia or other cancers that produce bone marrow failure. Individuals with

Fanconi anemia may pursue bone marrow transplantation, but so far there has been no consistently effective treatment.

Molly had Fanconi Anemia Type C. Although there are five subtypes of Fanconi Anemia, it is only Type C that occurs with increased frequency among individuals with Ashkenazi Jewish ancestry. Molly's mother, Lisa, and her father, Jack, both descended from Ashkenazi Jews. It is estimated that approximately one in eighty-nine Ashkenazi Jewish individuals is a carrier for Fanconi Anemia Type C. Parents of an affected child have a 25 percent chance in each future pregnancy of having another child with Fanconi anemia because they are both carriers of the disease.

Molly was facing progressive bone marrow failure unless she was given radiation treatments to kill the diseased cells. Those treatments would have to be followed by a bone marrow transplant to replace the diseased cells with healthy ones from a donor. It is a risky procedure, particularly for a child.

At the time of Molly's birth, the success rate of a bone marrow transplant from an unrelated donor was 18 percent. But the success rate for transplants from a brother or sister was as high as 65 percent.

Molly was Jack and Lisa's first child so she had no siblings. That's where our clinic and its ability to perform preimplantation genetic testing joined Molly's treatment program. This would mark the first time in history that medicine combined the sciences of in vitro fertilization, stem cells, and genetic testing. The goal was to provide Molly with the perfect bone marrow donor, which meant a healthy sibling with a tissue match.

Preimplantation Genetic Diagnosis then was new and controversial. Yet, we thought it would greatly increase the odds that Molly's parents would have another child who would be both healthy and a suitable bone marrow donor for her.

PGD involves testing multiple embryos created by in vitro fertilization. In Molly's case, we tested a single cell from each embryo to see if it possessed the genetic mutations that cause Fanconi anemia. Those embryos with the mutations were counted out while the healthy embryos were then tested to see if the tissue type was a match for Molly.

Unfortunately, Jack and Lisa had endured four costly IVF attempts without success at another clinic before they came to us. Even worse, Molly's health was rapidly declining because of leukemia. Her doctors concluded at one point that time was running out and Molly could not be saved.

"How do you explain to a five-year-old she's going to die?" her mother asked.

Lisa and Jack Nash came to our clinic as their last hope because we had developed a process for stabilizing an embryo—growing it longer in the Petri dish and giving it a much better chance of implanting. After coming to us, Lisa went through her fifth in vitro in one year. But this time, she produced twenty-four eggs, almost three times as many as she had previously. Half of the embryos were free of the FA mutations. Only one had no traces of FA and was also a bone marrow match.

We transferred that single "perfect" embryo to Lisa's womb on December 15, 1999. On Christmas Eve, a very weak Molly was undergoing a blood transfusion when my office called her mother to tell her she was pregnant. That was a joyful moment, though it was not an easy pregnancy for Lisa. She was also very concerned because Molly's health declined badly in the months that followed. Her parents and doctors weren't always sure the little girl would live long enough to receive her sibling's healthy bone marrow.

On August 29, 2000, Lisa Nash gave birth to a beautiful and healthy baby boy. She named him Adam. Blood from Adam's umbilical cord was harvested. Then, Molly's diseased bone marrow was destroyed by chemotherapy and radiation. After that was accomplished, the bone marrow transplant was performed. That final part of the life-saving process took all of twenty-five minutes.

Still, the fight was not over for Molly. She underwent a tortuous recovery. But she survived. She still faces many health challenges because of her genetic condition. We are hopeful, though, that one day, her parents may allow a grown-up Molly and her brother, Adam, to read the headlines from newspapers around the world: "Test-Tube Baby Born to Save His Sister."

❁ GENETICS 101

Chromosomes are the structures that contain our genes, which are the blueprints for our body. Each chromosome looks like a stick and contains one thousand or more genes. Genes are like the lines of a bar code along the stick (chromosome) with an estimated twenty-five thousand genes in total. There are twenty-three pairs of chromosomes, for a total of forty-six chromosomes, in each cell of the human body; twenty-three chromosomes originally come from the mother's egg and twenty-three chromosomes from the father's sperm. The chromosomes are numbered 1 through 22 by size with chromosome 1 being the largest; the twenty-third pair is the gender chromosomes, two X chromosomes for females or one X chromosome and one Y chromosome for males.

Human embryos can have the wrong number of chromosomes, which is called aneuploidy. This is by far the most common genetic problem occurring in embryos.

Less commonly, one gene (one line on the bar code) can be abnormal causing a gene change (mutation). Once the gene and mutation have been identified, specific PGD techniques are employed to determine which embryos are affected with the mutations and which embryos are unaffected and available for transfer to the uterus.

❀ *PGD—A REPRODUCTIVE OPTION*

Currently, I am the CCRM Scientific Director working from our flagship clinic outside Denver, but I once worked with babies and children who were born with genetic disorders. The suffering I saw inspired me to work as a reproductive geneticist. Now, instead of being reactive, I can be proactive.

Preimplantation Genetic Diagnosis is a reproductive option. It gives women a choice to begin their pregnancy with an unaffected embryo. They did not have this option before and that alone is a major advance. PGD allows us to go into the laboratory and examine early embryos for inherited genetic diseases and chromosomal disorders even before a patient becomes pregnant.

PGD was first developed in the early 1990s to screen embryos for X-linked diseases. The technology has expanded to detect countless other genetic disorders and chromosome abnormalities such as cystic fibrosis, Tay-Sachs disease, Down's syndrome, and muscular dystrophy. We can now even screen all twenty-three pairs of human chromosomes on a single cell.

The aim of PGD is to reduce the likelihood of implantation failure, miscarriage, and birth defects by only transferring embryos that are either not affected with a genetic disorder or that have the correct number of chromosomes. PGD is currently the only method available to determine an embryo's genetic or chromosomal status prior to pregnancy.

❀ THE PGD PROCESS

Of all of the reproductive diagnostic options available, PGD is the *only* technique that can be performed before pregnancy. This diagnosis is used to obtain genetic and/or chromosomal information from the oocyte (egg), day three embryo (cleavage stage), or day five embryo (blastocyst), thereby providing an alternative reproductive option for many patients.

The first step, a consultation with a genetic counselor, includes a review of family medical history and construction of a pedigree or family tree. For close relatives, it is important to note instances of such things as infertility, miscarriage, childhood or infant deaths, and individuals with congenital malformations or birth defects. A list of any hereditary diseases in either partner's family also is compiled.

All of this information helps us look for diseases or syndromes, both hereditary and acquired. Your genetic counselor is then able to assess whether there is a risk for having a child with an inherited disorder carried by a family member.

Genetic carrier screening is recommended to search for inherited mutations that are typically neutral if they occur on just one chromosome of the pair. However, if both partners are identified as carriers for a specific disorder, their chance of having an affected child is 25 percent. This is called recessive inheritance.

There is also dominant inheritance in which the parents can have clinical symptoms and in this form their chance of having an affected child is 50 percent. With both cases of inheritance, PGD is an option to identify unaffected embryos for transfer in an IVF cycle to establish a clinical healthy pregnancy. The diagnostic information provided by genetic carrier screening allows couples to make better-informed reproductive choices.

● *PGD FOR SINGLE GENE DISORDERS*

This form of PGD is a genetic test offered to couples with a family history of a single gene disorder. Five of the most common disorders for which PGD is used include cystic fibrosis, beta-thalassemia, myotonic dystrophy, Huntington disease, and fragile X syndrome. PGD is typically more than 95 percent effective at detecting embryos containing the mutation that will lead to the genetic disorder.

This allows couples to begin a pregnancy with an embryo not affected by the disease carried by the parents. It also means they don't have to make a decision to therapeutically abort an affected pregnancy. PGD for single gene disorders has been performed for hundreds of different disorders, both with dominant and recessive forms of inheritance.

● *BIOPSY FROM AN EGG*

The egg divides its chromosomes in half just before ovulation by ejecting twenty-three chromosomes out of the egg in a structure called the polar body. There are two polar bodies discarded by the egg during maturation and fertilization and they can be analyzed without removing a cell from the subsequent embryo. These structures will completely degrade within a day after fertilization.

The genetic information of both polar bodies is complementary to the chromosomes remaining inside the egg. The polar bodies tell us if the egg is genetically unaffected. Polar body PGD is most commonly used in situations where few eggs are collected following ovarian stimulation.

Because polar body PGD does not involve biopsy of the embryo itself, it is valuable for individuals with moral or religious conflicts to traditional PGD or for those living in countries with embryo protection laws. The main disadvantage of polar body PGD is that it provides information only about the egg's genetic contribution to the embryo.

● *BIOPSY FROM A DAY THREE EMBRYO*

The first embryo biopsy for PGD can be performed at the cleavage stage or day three of embryonic development, when the embryo contains six to eight cells. On the morning of day three, one blastomere (embryo cell) is biopsied from the embryo for genetic or chromosomal analysis. In contrast to polar body PGD, this method provides information about both the sperm and egg's genetic contribution to the embryo.

● *BIOPSY FROM A DAY FIVE EMBRYO*

Blastocyst PGD, which is the method we use at CCRM, is performed approximately five days after fertilization when the embryo has differentiated into two cell lines, the inner cell mass (ICM) and the trophoectoderm. The ICM will develop into the fetus, and the trophectoderm cells will become the placenta if a clinical pregnancy is established. Cells from the trophectoderm (precursor placenta) are removed without touching the future fetal cells (inner cell mass). Immediately following biopsy, the blastocysts are cryopreserved during the time of genetic analysis.

❋ *PGD FOR CHROMOSOMAL ABNORMALITIES*

Structural chromosomal abnormalities arise when breaks occur in the chromosomes, causing segments to be lost, rotated, or even inserted into neighboring chromosomes. An individual with a structural chromosomal abnormality is often completely normal because he or she has the correct (balanced) amount of chromosomal material; it is just rearranged. However, if patients are carriers for structural chromosomal abnormalities, they have significantly higher chances of producing eggs or sperm with the incorrect number of chromosomes. This will result in embryos that have chromosomal errors explaining their infertility.

A blood test called a chromosome analysis or karyotype can determine which patients have these structural abnormalities. For such couples, the use of PGD as a tool for selecting embryos prior to transfer can help to ensure the normal number of chromosomes in their offspring.

❋ *PGD FOR ANEUPLOIDY*

Numerical chromosomal abnormalities, or aneuploidy, is caused during cell division when the chromosomes do not separate correctly resulting in too many or too few chromosomes in the resulting cells. When aneuploidy occurs during the formation of the egg or sperm, the resulting embryo will also have the incorrect number of chromosomes.

Chromosome abnormalities have a major impact on embryo viability and are one of the principal causes of failed IVF attempts. Unfortunately, aneuploidy increases with maternal age. In women between the ages of thirty-eight and thirty-nine years, aneuploidy can occur in 50–60 percent of eggs, and this increases to 80 percent and more for women over forty years of age.

The increase in egg aneuploidy seen with advancing maternal age may be associated with the age-related decline in IVF success rates. The lethality of chromosome aneuploidy is also evident in the fact that the majority (70 percent) of naturally conceived pregnancies resulting in spontaneous first trimester miscarriages have an abnormal number of chromosomes.

Unfortunately, we cannot detect chromosome abnormalities by simply looking at the appearance of the embryo under the microscope. Therefore, it is reasonable to screen embryos for aneuploidy (incorrect chromosome number) to identify and transfer only those with the correct chromosome constitution. This leads to increased pregnancy rates and decreased risks of miscarriage.

For these reasons, PGD was developed to screen for chromosome aneuploidy or comprehensive chromosome screening (CCS), and over the last five years, it has gained popularity due to the significant improvements in clinical outcomes for infertility patients.

Potential patients include:

1. Advanced maternal age: >35 years

Advanced maternal age is a major risk factor for oocyte chromosomal aneuploidy. One of the most common disorders is Down syndrome, which is caused by an extra chromosome 21, and, thus, its alternative name is trisomy 21. Errors of other chromosomes lead to embryos that often don't implant or 100 percent of the time result in early pregnancy loss. PGD for chromosome aneuploidy aims to prevent these situations in a pregnancy by selecting only chromosomally normal embryos for transfer.

2. Recurrent miscarriages

Recurrent miscarriage is typically defined as three or more consecutive miscarriages of fetuses between six and twenty weeks' gestation. These patients are estimated to include 0.3–2 percent of the population. Most of these miscarriages are a result of advanced maternal age with a high proportion resulting from chromosome aneuploidy. PGD can be used to identify chromosomally normal embryos and avoid continued miscarriages.

3. Recurrent IVF or implantation failure

Recurrent IVF failure is most often defined as three or more subsequent failed IVF attempts (implantation failure) or failure after the replacement of more than 10 embryos. The most commonly considered causes of recurrent implantation failure are endometrial abnormalities, advanced reproductive age, and presence of endometrial polyps. Though the potential causes of recurrent IVF failure are diverse, a correlation exists between a higher frequency of chromosomal aneuploidies and failed IVF attempts.

PGD for chromosome aneuploidy is extremely valuable because euploid embryos (those with the correct number of chromosomes) can be identified and transferred to establish a pregnancy. In addition, PGD also provides patients with diagnostic information regarding their IVF embryos. For example, if no normal embryos are found in one cycle of IVF, the chances of having all abnormal embryos in a subsequent cycle are significantly increased. In this situation, oocyte donation may be recommended.

❁ COMPREHENSIVE CHROMOSOME SCREENING PLATFORMS

As pioneers of blastocyst CCS, the team at CCRM has implemented technology changes over the years allowing for developments in the laboratory to be tested and transferred into clinical practice. These have provided fantastic benefits to patients including faster turnaround time for results and added cost effectiveness.

The latest breakthrough in CCS technology is the introduction of next generation sequencing (NGS). Other molecular techniques we have utilized since our pioneering first blastocyst CCS case in 2007, include metaphase comparative genomic hybridization (CGH), single nucleotide polymorphism (SNP) microarray, array CGH and PCR-based detection. All of these different technologies have been equally effective in testing for all twenty-three chromosomes.

Our clinic has been involved also in conducting a well-controlled randomized prospective trial using CCS to determine the clinical efficiency of the procedure for women of advanced maternal age (greater than thirty-five years). Our novel CCS strategy involves biopsy of trophectoderm (precursor placental) cells from the embryo at the blastocyst stage (day five or six of embryonic development). The blastocysts biopsied for PGD need to be cryopreserved (frozen) while genetic analysis takes place. This is an advantage because better pregnancy and delivery outcomes have been observed with a frozen embryo transfer (as explained in the previous chapter).

Our results to date indicate striking improvements in live birth rates for women of advanced maternal age, suggesting that the clinical efficiency of PGD for chromosome aneuploidy may have been reached with our novel CCS strategy. To determine these results, we compared advanced maternal age patients who underwent day five fresh embryo transfers without CCS (control group) to our study of advanced maternal age CCS patients with a frozen embryo transfer.

Embryos chosen for transfer in the control group were selected on the basis of traditional appearance (morphological analysis), while those from the CCS study were selected on the basis of CCS testing with only chromosomally normal blastocysts. The control cycles had the same clinical parameters (maternal age, day three FSH level, number of oocytes retrieved, fertilization rate, number of blastocysts produced, and number of previous unsuccessful IVF attempts).

The probability of one individual blastocyst implanting and resulting in a live birth after transfer was significantly increased in the study group; 74 percent with CCS compared to 51 percent without CCS screening. Additionally, the miscarriage rate was significantly reduced in the study group; 3 percent for the CCS tested pregnancies compared to nearly 21 percent without CCS screening.

On the basis of this data, we suggest that this CCS strategy involving blastocyst biopsy, comprehensive chromosome screening and vitrification with a frozen embryo transfer significantly improves clinical outcome for advanced maternal age patients, with increased implantation, decreased miscarriage and higher live birth rates. These excellent clinical outcomes also allow for the implementation of routine single embryo transfer for advanced maternal age patients.

I thought it would be helpful if our patient Tracey offered her perspective on the value of PGD in her case:

> In 1991, at the age of twenty, I was diagnosed with an unknown fertility issue while serving as a Navy Corpsman. Because I was active duty military, I was able to get a wide range of tests and treatments, none of which, unfortunately, led to a successful pregnancy. During my time in the Navy, we tried treatments including ovarian drilling and artificial insemination, but everything failed.
>
> My Navy doctors said IVF was my only hope. I left the Navy in 2000, and moved to Colorado to be with my husband, Michael. I was referred to a doctor now working with Doctor Schoolcraft's clinic. Once she'd read the massive medical history of my fight for fertility, she decided to start at ground zero with blood work and sperm samples. We again tried artificial insemination a couple of times without success.
>
> In 2001, we tried another IVF cycle with another doctor who was in the network covered by my insurance. It was unsuccessful and ended in hyperstimulation. At this time, we decided to cover the cost the best we could and go back to the doctor at CCRM. After another round of tests showed everything to be normal, she suggested genetic

diagnosis. We agreed because we were desperate for an explanation.

The genetic diagnosis found that my husband had no issues, but I had a chromosome rearrangement that would make it very difficult to conceive. This was devastating news, but after all of the heartache we had already been through, we decided to continue and hope for the best.

Our first round of IVF went forward and we retrieved four oocytes. Testing showed that only one was suitable for transfer, which we did. This attempt was also, sadly, unsuccessful. We decided to give my body, and our weary spirits, a rest.

In 2003, we were ready to try again. Then, my husband, who is a U.S. Marine Reservist, was informed of his deployment to Iraq. We were determined to continue our efforts so we froze Michael's sperm before he left. I then began round two, only this time without him there to support me. Still, the day of retrieval was fantastic!

We had numerous oocytes, several of which were fertilized through ICSI and sent off for PGD. The results came back and four or five embryos were suitable genetically and of good quality. While this may not seem like many to most people, it was a treasure trove for us.

The day of transfer came, with my husband still in Iraq. I was sick as a dog. There were suggestions that I should wait until I felt better, but I had waited so long already I wanted to get it done. Two embryos were transferred that day and they both grew well, initially. We lost one of them in the second trimester, with my husband still deployed.

Finally, on December 17, 2004 our beautiful baby boy, Rhys, was born. Unfortunately, he has the same chromosome rearrangement as I do, which means that he too may face challenges when he wants to become a parent.

About two years later, Michael had returned safely from Iraq so we attempted the whole process again to try to have another child. The doctors at CCRM transferred the last two of my frozen embryos. This round went much more smoothly and I became pregnant. This time only one of the embryos implanted, but on March 6, 2007 our second beautiful baby boy, Owyn, was born. He has normal chromosomes.

My husband and I highly recommend getting genetic diagnosis done early on versus waiting as long as we did. Knowing whether or not you have a genetic issue can save time, money, and heartbreak. The testing helped us understand our challenges and gave us peace of mind so we could do what we had to do to have a family.

CHAPTER FIVE

GESTATIONAL CARRIERS

BY DOCTOR JIM TONER, M.D. PH.D.

Susan and Brian met in college, and married four years later. For the first few years of their marriage, they concentrated on establishing a home and their careers. In their late twenties, they began to try to have children, but after two years, Susan was still not pregnant. They reached out to Susan's obstetrician for help, and initial testing showed Brian had a low sperm count. The obstetrician referred them to a fertility specialist for further consultation.

Testing for genetic and hormonal causes of low sperm count revealed nothing. Their fertility doctor said Brian's sperm count was too low for intrauterine inseminations to work, but there were enough sperm for in vitro fertilization (IVF).

Susan and Brian decided to do IVF, and their attempt worked. They were thrilled. Since Susan was still under thirty and had good ovarian reserve, she responded well to the medications. Twenty-four eggs were retrieved, which is a very good result.

A single sperm was injected into each of the twenty mature eggs, and fifteen of them fertilized. After five days, a single fully developed blastocyst embryo was transferred to the uterine cavity, and four others were frozen for use later on. Susan and Brian were thrilled when they got the call, ten days later, that the pregnancy test was positive.

Susan's pregnancy was normal from start to finish, and their daughter Mary was born on her due date after a normal labor. However, a few hours after the delivery, Susan started to have heavy bleeding, and the usual medicines used to stop it did not work for her.

Susan had to be taken to the operating room to control the bleeding, and in the end, she needed a hysterectomy. The joy of Mary's birth was tempered by the realization that Susan could no longer carry and deliver a pregnancy, though she still had four embryos in the freezer.

When Susan and Brian first saw me after the delivery about a year later, I was unaware that she had had a postpartum hemorrhage requiring hysterectomy. I thought they were coming back for baby number two, using the frozen embryos in Susan's uterus. But without a uterus, Susan would need another woman to carry her pregnancy. From a medical standpoint, this is easy to do. The embryo doesn't care which uterus it grows in, and preparing the uterus for pregnancy is the same in every woman.

Susan and Brian had done their research and they wanted to use a gestational carrier, whom they'd found through a local attorney. The recommended carrier passed all the screening tests, and after a single blastocyst embryo was transferred into her uterus, she became pregnant.

While using a gestational carrier is easy from a medical standpoint, it is complicated in many other ways. Legal, psychological, social, and financial issues arise, and it takes a team to navigate them. In fact, the laws on gestational carriers are not the same in every nation.

The use of surrogates has become widely accepted in the United States, but it remains a very controversial issue in other countries, including France where it was declared illegal in 1994. Until recently, even the children of French couples who were born abroad to a surrogate mother were not recognized as citizens in their parents' country.

The staff of the Colorado Center for Reproductive Medicine encountered this situation when Denver native Sarah became a patient. She was forty years old, married and living in Paris with her French husband, Eric Landot, when she came to the CCRM headquarters outside Denver in hopes of having a child through in vitro fertilization. Sarah and Eric had been trying for a year to conceive a child without luck. She'd had one miscarriage and one ectopic pregnancy.

During CCRM's standard and quite thorough battery of tests in preparation for IVF, it was discovered that Sarah had endometrial cancer. This was not welcome news, of course, but this early detection may well have saved Sarah's life. Understandably, she was overwhelmed with grief at first. She feared the cancer endangered not only her health, but also her ability to have a child.

The CCRM staff explained to Sarah that her cancer was very treatable and there were still options available to her for having children with her husband. Doctor Schoolcraft and his team in Colorado felt confident that they could help Sarah once she had conquered cancer.

Still, no one could have predicted back then that one day she and Eric would have three children together—all born with the help of different gestational carriers. That is exactly what occurred. Sarah and her first gestational carrier, Aimee Melton, even wrote a book about it, in French. The title translates in English to "When We Only Have Love."

I thought Sarah, who was in the pharmaceutical industry for many years, would be another excellent example and guide on the subject of gestational carriers because of her depth of experience and all of the research she has done on this topic.

When she was struggling in her efforts to have a child, her fertility doctor in France learned that she was planning on a visit to her family in Denver. He told her that her hometown had "the best fertility center in the world," and that is what brought Sarah to CCRM. She had actually known Doctor Schoolcraft earlier in his career when he served as her mother's gynecologist before opening his own center for reproductive medicine.

Even though fertility treatment was more expensive in the U.S. because France has socialized medicine, Sarah knew that she was closing in on her fortieth birthday and her fertility window was diminishing. The endometrial cancer diagnosis further complicated and delayed her efforts to have a child, but Doctor Schoolcraft assured her that the cancer had been detected in an early stage. Her uterus would have to be removed but her ovaries would remain intact. She still could make eggs that could be fertilized with her husband's sperm to create their own biological child.

Doctor Schoolcraft said Sarah could have an IVF procedure prior to the hysterectomy and her embryos could be frozen for later use. Because she did not need extensive cancer treatments following her hysterectomy, Sarah was able to have another successful IVF when it was done. After six months of treatments and recovery, Sarah was cancer free. She then renewed her efforts to have a child, using her frozen embryos and a gestational carrier.

❀ *THE HISTORY OF GESTATIONAL CARRIERS*

The first use of a gestational carrier involved inseminating a volunteer, or paid surrogate or carrier, with the sperm of the intended father. This approach is known as "traditional surrogacy." Problems arose when some surrogates decided they wanted to keep the babies. Lawyers were hired and legal battles raged. Couples who just wanted to have a baby were dragged into costly and time-consuming court cases.

The traditional surrogates who decided to keep the babies had strong cases because their eggs were used. The children they'd borne carried not only the fathers' genes but theirs as well. Due to these legal challenges from traditional surrogates, some states adopted strict legislation that either does not recognize these contractual agreements or prohibits them. Other states have either no provisions against surrogacy or have ruled in favor of cases concerning surrogacy agreements.

Consequently, the use of traditional surrogacy is now rare. Instead of the egg coming from the couple's surrogate, most now use eggs from another woman, either the aspiring mother or a donor. With this arrangement, there have been no claims, let alone successful legal challenges by carriers trying to keep the child as theirs, because the baby is not theirs biologically.

❀ *WHO MIGHT NEED A GESTATIONAL CARRIER?*

A gestational carrier is appropriate for those:
- who've had hysterectomies
- with uterine problems that cannot be fixed. Sometimes women are born with abnormally small uteri that cannot carry a child to term. Other women have

uteri deformed by prior fibroid surgery, or have had uterine artery embolization.

- with conditions such as heart disease or Turner's syndrome that make pregnancy a high risk for them
- who've experienced multiple miscarriages
- who've had multiple IVF failures
- who are men without female partners

There are also situations where use of a carrier is not appropriate. It's not typically recommended for women who are themselves able to carry their pregnancy but would prefer not to do so. Pregnancy is not risk free, and it is ethically unsound to have another person bear the risk of another without a sufficient need.

❀ *THE PROCESS*

The process for finding and employing a gestational carrier has its challenges, but the growing popularity of this option has made it less stressful in recent years. There are several parts to the process. They include:

- identifying a possible carrier through your fertility specialist, a personal referral or an agency specializing in representing them
- evaluating her suitability to be a carrier
- engaging an attorney experienced in reproductive law in your state
- going through the treatment cycle of the gestational carrier

Let's look at each step in the process. Identifying a carrier can be one of the more difficult steps. Sometimes clinics will provide a list of carriers they recommend. In other cases, lawyers who specialize in this area have a list of carriers. There are also a growing number of gestational carrier agencies that have recruited and done screening of potential carriers. The internet is one way to find these agencies, but a personal reference is probably the best route. Make sure you thoroughly check out the reputations of any agency you use—and that of any gestational carrier you select.

Your screening of the carrier should include criminal and financial background checks. A personal reference or two, or even three is also highly recommended. While an initial medical screening may be done by the agency, the final approval of a carrier lies with the treating clinic.

❁ *EVALUATING THE CARRIER'S SUITABILITY*

Successful carriers must first and foremost be able to successfully carry and deliver a pregnancy. To evaluate this aspect, her prior obstetric history will be carefully reviewed, and well as her overall health. The condition of her uterus will be evaluated to make sure it's normal, and her cervix will be evaluated to be sure the embryo transfer catheter can pass easily.

But there are other important aspects as well:

- Suitable age for safe pregnancy is normally between twenty-one and forty.
- Healthy weight—Body mass index should be between twenty and thirty.

- Stable lifestyle—No financial hardship or government assistance programs.

- Healthy lifestyle—No smoking, drinking, or recreational drug use.

- Stable mental health—This will be evaluated by mental health professionals.

One of the most important first steps is for the intended parents to meet with a mental health professional to discuss the proposed GC arrangement. This includes the reasons for choosing to pursue this treatment, as well as the relationship and expectations they have with the gestational carrier.

In a later meeting, the mental health professional will meet with the gestational carrier and her partner. Based on the results of meetings with the clinical mental health professional, additional appointments or psychological testing may be required. Mental health professionals used for evaluation of the intended parents and the gestational carrier need to be experienced with "third-party reproduction," including being familiar with the appropriate professional guidelines.

The visits will explore critical topics in depth, including communication and expectations of all parties before, during, and after birth; consideration of pregnancy with multiples (twins/triplets), and potential fetal-reduction or abortion for medical reasons. Later in this chapter, a gestational carrier will explain why it is so important to discuss these matters thoroughly so that the aspiring parents and the woman bearing the child understand and respect each other. She will note that in her experience as a gestational carrier, two of the potential couples she worked with were open to personally bonding with her while a third couple preferred to keep things more businesslike.

Every relationship between a gestational carrier and her client or clients is unique. You should have a good sense of what your relationship will be like before you sign on the dotted line. You will probably want to establish boundaries or ground rules on things like how much input the aspiring parent or parents have during the pregnancy of the gestational carrier so resentments and frustrations are kept to a minimum. Additional topics of discussion may also include consideration of future disclosure to the intended parents and gestational carrier's children, and conflict resolution in the event of unforeseen conflicts.

Not all prospective surrogates are approved, even those with the best of intentions. The standards for a gestational carrier are high because of the emotional and psychological issues involved in carrying someone else's child, and because the steps to complete the process can be complex.

The purpose of all of these evaluations is to protect the interests and emotional health of both the gestational carrier (and where relevant, her partner and children) and the intended parents. Even with them, no one can predict future behavior. The appointments are mandatory, and a letter from the clinical mental health professional must be received by the clinic before treatments proceed.

❁ *ENGAGING SERVICES OF AN ATTORNEY*

Your fertility clinic staff can't provide legal advice, so the gestational carrier and intended parent(s) must have ongoing legal counsel by an appropriately qualified legal practitioner who is experienced with third-party reproduction and licensed to practice in the relevant state or states, or in the event of an international arrangement, the intended parent(s)' home country.

A legal contract prepared by such counsel must be in place prior to commencement of an IVF cycle and any subsequent embryo transfer. This contract should contain information designating the roles of all participants involved with respect to parental rights. Additionally, the contract should address all relevant issues, including but not limited to: the amount, timing, and escrowing of compensation (if any), insurance coverage for the pregnancy and offspring, medical care and decision making, and parental rights.

All parties involved should be aware that state law pertaining to gestational carrier arrangements and parental rights resulting from such arrangements differ from state to state. Each participant, including the gestational carrier and her spouse (if there is one), as well as the intended parent(s), should retain and be represented by a lawyer with relevant experience in family law and gestational carrier contracts.

The participants will have to take legal actions to allow the intended parent or parents to be recognized as the legal parents and hopefully entered on the child's birth certificate. This may require a court petition prior to the birth of the baby and should be handled by an experienced reproductive technology law attorney.

A letter of clearance from the respective attorneys will be required prior to treatment with a statement that a valid legal contract has been prepared, negotiated, and properly entered into to the satisfaction of the intended parent(s) and the gestational carrier (and any spouse), that the gestational carrier and the parents have had independent counsel, and all participants understand and assume the risks of proceeding. Depending on state laws, a birth order may be obtained to place the intended parents' name on the birth certificate, while in other states adoption proceedings may be required.

❀ TREATMENT CYCLE OF GESTATIONAL CARRIER

A gestational carrier cycle usually involves these steps:

- Oral Contraceptive Pills (OCP's) for one to three weeks
- Pituitary suppression using GnRH agonists or antagonists (e.g., leuprolide, Lupron®, ganirelix, Cetrotide®)
- Development of mature endometrium (uterine lining) using transdermal estradiol (Vivelle Patches®, Estraderm®), vaginal estradiol, or injectable estradiol
- Progesterone for luteal phase support. Progesterone may be given as an injection or vaginal preparation (Prometrium®, Crinone®, Endometrin®).
- Transfer of the embryo(s) back into the uterus
- Pregnancy test

❀ FINANCIAL CONSIDERATIONS

When working with a gestational carrier agency, please ensure that any fees paid to the agency and the gestational carrier are placed in a legitimate escrow account for the safety of all parties. Be aware that these agencies are not regulated. The agency should not have the ability to withdraw any money from the escrow account. It is of the upmost importance that intended parents verify the legitimacy of the services that are to be provided. Intended parents also should make certain that the agency and its escrow account are insured.

Insurance coverage for any pregnancy and possible complications can be expensive. Usually, the intended parent(s)' insurance policy will not cover the pregnancy or any associated treatments when a gestational carrier is involved. The gestational carrier's insurance may not cover the pregnancy expenses when the baby will not be the legal child of the gestational carrier.

The intended parents should also make sure their policy will cover newborn care, and they should be aware of any conditions that may apply to obtaining such coverage. Several insurance agencies offer insurance coverage for gestational carriers' medical expenses. It is your responsibility as a participant to a gestational carrier arrangement to ensure there is adequate insurance coverage in place for the pregnancy, delivery, and child before proceeding with treatments.

Your clinic and its staff cannot provide you with advice regarding insurance coverage other than the contact information for various insurance agencies. The clinic will require that you confirm through your attorneys within your letter of clearance that you have obtained adequate insurance coverage.

Costs are high, and are most often not covered by insurance. The table below indicates average costs of the essential services, totaling about sixty thousand dollars. This does not include the costs of obtaining the eggs initially, which involve routine IVF or Donor Egg therapy. Fees can vary state to state and agency to agency.

Carrier Services	Approximate Cost
Agency fee	$20,000—$35,000
Carrier fee	$30,000—$45,000
Mental Health services	$4,000
Legal services	$6,000

❀ SARAH'S STORY CONTINUED

Because using surrogates and gestational carriers is illegal in France, Sarah and Eric had many discussions before they decided to find a gestational carrier in the United States. They were fortunate that a family friend, Aimee Melton, volunteered. Already a mother of two children, Aimee had worked as a midwife and had a longtime interest in surrogacy.

Sarah and Eric agreed to pay her a fee and cover her medical expenses and legal and insurance costs. Their total expenditures were about twenty-two thousand dollars. Initially, two embryos were transferred and neither survived. On the second attempt, two embryos were transferred and one survived. Aimee delivered a boy, who was named Oscar, in 2011.

Sarah and Aimee bonded as sisters during the pregnancy. Sarah believes that Aimee "fell from heaven." Aimee felt that Sarah and Eric deserved to have a child together and serving as their gestational carrier proved to be "an amazing experience." Their book published in France is built around a diary that Aimee kept of their ultimately successful partnership.

Aimee and Sarah had met through Aimee's friendship with Sarah's sister, Jessica. After seeing how well Aimee and Sarah got along during the creation of Oscar, Jessica stepped up and offered to carry another child for Sarah and Eric.

Jessica also works as a midwife.

Two embryos were transferred into her uterus initially. She became pregnant with twins, but miscarried both. Two more embryos were transferred after she recovered from the miscarriages. No pregnancy resulted, so two more embryos were transferred. Jessica became pregnant with one child and delivered Sarah and Eric's daughter, Vivianne, in 2013.

The couple, who still live in France, were pleased on July 3, 2015 when the country's highest court granted legal recognition to surrogate children. But they were ecstatic over another event that very same day—the birth of their son, Kennan. This time the gestational carrier was another Colorado woman, Jessica, who has two children of her own and works as a doula to assist women during pregnancies and births.

Sarah Levine said the three women who carried her children brought incredible love to her family. She considers them "womb sisters."

"Without them, I would have had a very different relationship with my husband and a different life," Sarah says. "The thought of staying home instead of working hadn't entered my mind, but after my miscarriage I said that if I was ever lucky enough to have a child there was no way I'd go back to work. I am incredibly blessed to have these three children—and to have these three women as my sisters."

❀ A GESTATIONAL CARRIER'S VIEWPOINT

I thought it would be helpful to share with you the viewpoint of a gestational carrier with varied experiences. This woman, Nancy, works at our clinic in Colorado as a clinical lab assistant. She initially came to us as a patient five years before joining our staff. She had been struggling to get pregnant for the previous three years.

The team at CCRM helped her have her first child through IVF. She became pregnant with her second "miracle baby" when her first was just five months old. She has felt blessed by her children and empathetic to other women who were struggling with infertility.

She and her husband were not looking to have more children of their own, but Nancy thought it would be a great experience to serve as a gestational carrier.

As a woman who had struggled to have her first child, Nancy understood what a life-changing gift she could provide to another couple.

In hopes of doing that, she signed up with a California agency that pairs gestational carriers with clients. Her second pairing, which was successful, was with a San Francisco couple. The wife had become pregnant several times, but had never been able to carry a baby to delivery. An embryo transfer was performed and Nancy became pregnant with twins on the first try. She had a very low-stress pregnancy and the twins were delivered at thirty-nine weeks. Both weighed about five pounds.

"It was an awesome journey, and I loved being pregnant," she said. "I never experienced morning sickness or any of the common problems associated with multiple births."

Nancy worked with three couples altogether before retiring as a gestational carrier. She bonded closely with two of them but not the other, which she understood was the result of cultural differences, and didn't take it personally. Nancy's early struggles with infertility made her all the more appreciative of motherhood once she had children of her own. It was a gift she wanted to share with others.

"I had struggled before I became a mother and I wanted other women to know what it was like to be a mom," she said. "Seeing their excitement when their babies were born was the best part for me.

"My advice for other gestational carriers would be to go into this for the journey, not the actual pregnancy. In the end, even if it doesn't work out, it will still be an important part of your life and the lives of the intended parents. I'd also suggest that you give the intended parents every opportunity to treat this as their pregnancy, not yours. Every woman has different ideas of what it will be like, so allow the intended mother to fulfill her vision of pregnancy by following her suggestions and advice as much as possible. Understand that the couple has been waiting a long time and this probably isn't the route to parenthood they had envisioned originally, so go with the flow. And keep in mind that this experience may not turn out exactly as expected."

❀ HOLDING ON TO HOPE

Nancy and other gestational carriers are a great source of hope for those women who cannot carry their own babies full term. I encourage our patients to remain open to the options that modern science provides them. Even those who have had one failure after another shouldn't give up hope because reproductive science continues to make rapid advancements.

Fiona was one of those patients who went through repeated failures but refused to give up. A resident of rural Maine, she was nearing forty years old when she came to us. She had been married four years to a man she'd dated for three years before their marriage. She'd become pregnant in 2005, but they'd lost the baby during the delivery.

"I was in labor and everything was great and all of a sudden they couldn't find the heartbeat," Fiona recalled. "It didn't make sense that she would die. It never occurred to me that you could lose a baby like that."

Fiona and her husband went through all the stages of grief and shock. Like many couples who've lost a child, they felt alone and isolated because it was so difficult to talk about with other couples who'd never experienced such grief. After a long period of mourning, Fiona decided to try again, but she couldn't get pregnant. In 2008, they tried in vitro fertilization at a clinic in Maine. She became pregnant with a boy, who was, tragically, stillborn in the third trimester.

Losing another child in this manner was devastating, and a deciding factor in this couple's decision to explore other options. "The stress of infertility was a whole new level of emotional pain," she said. "My husband says he felt like we lost an entire decade of our lives because we were so focused on trying to have a baby. It was a dark period in our lives."

Though she recalls "being at the end of my rope," Fiona did not give up. She continued researching developments in reproductive science, including the use of comprehensive chromosome screening (CCS), which allows for selection of the healthiest embryos for transplant after IVF. Her research led her to the CCRM clinic in Denver.

"I was on an online forum to learn about IVF and there was a sub-forum for women going to CCRM. One woman made a chart of all those who were in a study they started in 2009 for those who'd gone through comprehensive chromosome screening. The chart showed the cycles and outcomes and the births. There were thirty women and the success rates were mind blowing because our local clinic had zero success rate for women my age," said Fiona.

Though it was a long way from their home in Maine, Fiona and her husband decided that Doctor Schoolcraft and his team offered their best hope of finally having a child. They flew to Denver. After undergoing tests and consultations, they decided to try IVF again, this time with CCS to screen for the healthiest embryos.

"They were able to give me perfect care and we got thirty eggs through IVF, it was like hitting the jackpot because it gave us so many embryos to work with," she said.

The embryos were screened and the healthiest were selected. Fiona followed the strict protocols designed to produce the optimum results. Because of her past experiences with failed pregnancies, they monitored her progress every step of the way.

This proved to be important because her pregnancy was complicated with severe velamentous cord insertion (VCI). In this abnormal condition, the umbilical cord doesn't insert into the middle of the placenta. Instead, it inserts into the fetal membranes, then travels within them to the placenta. This makes the cord vulnerable to rupture especially if they are near the cervix—a serious condition called *vasa previa*. This was the situation in Fiona's pregnancy and it can result in a stillbirth during early labor. She had to be closely monitored.

Fiona was required to spend six weeks on hospital bed rest and she stayed in the room closest to the operating room in case she needed an emergency C-section. Fortunately, Fiona was a very dedicated patient and thanks to her great attitude and top quality care, she was able to deliver a healthy son in 2010.

"The anxiety and fear I experienced during this pregnancy is difficult to put into words," she recalled. "There was never a question in my mind that I would never attempt to carry another baby because when our son was delivered alive and healthy, we were told that we were very lucky considering the severity of the complications. That was the final factor in deciding to use a gestational carrier with our next child."

Then, three years later, she came back to us in hopes of having one more child. She still had healthy frozen embryos in storage that could be used. Doctor Schoolcraft and Fiona were concerned about the complications she and her baby had endured as well as the fact that she'd had five pregnancies that were each made difficult by very different factors. For those reasons, Fiona decided to use a gestational carrier.

She worked with an agency that had provided gestational carriers to previous CCRM patients. Her carrier was twenty-seven years old, married with a supportive husband, and healthy children. This young woman had decided not to have any more children of her own. She had been carefully screened both by CCRM as well as the surrogate agency. We wanted to be certain she was the right candidate—both physically and emotionally—to carry a baby for Fiona.

"We had a great experience. She was very professional and really amazing," Fiona said. "It was just a completely different kind of experience. I was very happy not to have to go through the stress of a pregnancy and relive that fear of my baby dying again. Some women who can't carry a baby feel they are missing out using a surrogate, but I didn't feel that way at all. I was happy not to have the stress. To be at the birth and see my daughter born was wonderful. It was an even more joyful experience than I thought it would be."

The birth also proved to be joyful and highly emotional for the gestational carrier and her husband. In fact, her parents attended the birth so they could share in the beauty of the gift she provided Fiona and her husband.

The gestational carrier experience was a great one, but it was also a challenge to the finances of this couple. Because health insurance does not usually cover the IVF treatments or the many costs of gestational surrogacy, Fiona and her husband had to use their savings and borrow money to finance their efforts to overcome fertility challenges. They estimate their total expense over the years at around two hundred thousand dollars.

Even so, Fiona feels incredibly blessed to be the mother of two children after so many years of struggling with fertility challenges.

"I definitely would not have my daughter if not for frozen embryos, gestational carriers, and the technology that now exists," she said. "You find yourself with these challenges you never expected, but you can persevere and find different paths for having children. People can't believe I kept trying. Honestly, I never thought it would take so long or be so difficult. Someone told me that you'd never think of jumping off a three-story building until you're in one and it's on fire. You find the strength to do what you have to do."

CHAPTER SIX

EGG FREEZING AS FAMILY INSURANCE

BY DOCTOR WILLIAM SCHOOLCRAFT

Most of my days as a reproductive scientist and physician are spent consulting with and treating patients, doing research, or refining methods in our lab at the Colorado Center for Reproductive Medicine. It was an unusual experience, then, to find myself one evening in a large room full of women, surrounded by lively music, a stocked bar, and several rows of long tables offering every imaginable crafting option, from watercolors and coloring books to kits for making necklaces, string art, wallets, and pet collars.

The location was Upstairs Circus in the heart of Denver's downtown historic district. Although the setting was quite fun, it was, in fact, also part of my mission to help women who want to have children according to their own schedules.

The occasion was an attempt to spread the word on what I consider to be one of the most important developments for women's independence in the last forty years—a proven scientific method for safely freezing their eggs while young, and storing them until they are ready to have children.

I will describe this process, known as oocyte vitrification, or cryopreservation, in greater depth later. Originally developed for women facing cancer treatments likely to make them infertile, it involves stimulating a woman's ovaries with hormones to produce multiple eggs, retrieving the eggs from the ovaries and then cooling them to subzero temperatures to preserve them so that they can be thawed and fertilized when the woman is ready to conceive through in vitro fertilization (IVF).

This is a groundbreaking, liberating, and even historic development for all women who want to have children on their own schedules. The Denver event, held in March 2015, was our first effort to get the word out with an "Egg Freezing Party." These gatherings have become popular around the country. Despite its billing as a party, I wasn't sure just how much of a social occasion it would be.

We had run advertisements online and in the media inviting women interested in the egg freezing process. I knew most of them would have concerns and serious questions about their fertility. I wondered if they'd really want to do crafts too, but the Upstairs Circus staff had assured us that women enjoy working with their hands while they talk and share experiences.

Those assurances proved to be spot on. One by one, nearly forty women arrived. They dove in, selecting their crafting materials and joining those around them in conversation. Soon the room was alive with their conversations, comments, and laughter. They seemed to be enjoying the crafting projects and each other's company so much that I felt guilty interrupting their party to begin our presentation.

I shouldn't have worried. As my colleagues and I spoke and answered questions, the women continued to work on their crafting projects. All the while, they raised thoughtful questions and concerns. It was clear that most of them had already conducted a good deal of research and had carefully considered their options, desires, and dreams.

One woman asked if being on birth control for fifteen years would adversely impact the quality of her eggs. (Not usually.) Another wondered if the egg freezing process would harm her fertility. (No.) A third asked if the shots would make her highly emotional because of the hormones. (This does happen, but usually it's no worse than typical PMS, most say.)

Many of the participants were single, working women who were approaching or had reached their mid-thirties and were concerned about their declining fertility. "I just want to take the pressure off because every time I date someone, it feels like I have to decide right away whether he is going to be 'the one' or whether I should move on to someone else,'" said Gretchen.

Gretchen, a thirty-four-year-old who works in medical sales, had just re-entered the dating world. She said. "I'm getting older and I'm not in a relationship, but I always wanted to have kids. I know the risk of having them at an older age and I've experienced the emotional stress of the 'ticking clock' that makes you feel things have to be rushed along. I'm tired of that," she said.

Like most women, Gretchen doesn't want to rush into marriage just to have children. She would prefer to marry and spend a couple years settling into a secure relationship before starting a family. Freezing her eggs would give her time to "find Mr. Right rather than Mr. Right Now," as one of our patients described it.

Men have long been able to delay parenthood until after they've earned their degrees, established their careers and then decided, often in their thirties, to settle down and have a family. The science that allows women to safely freeze and store their eggs finally gives them the same freedom and opportunities. I'd say it's just as significant as the invention of birth control. Now, those who want to delay having children can freeze their eggs and relieve the pressure of worrying about declining fertility after their twenties.

Many women who have gone through the process say freezing their eggs helped them take control of their own destinies. One woman, interviewed in a Vogue magazine story on modern reproductive science, said freezing her healthy eggs is a way of "freeing myself from the tyranny of the expiration date." (Time to Chill? Egg-freezing Technology Offers Women a Chance to Extend Their Fertility Vogue.com, April 28, 2011)

Our patients often see it as an insurance policy for family planning in the future. You don't have to use the frozen eggs if your own are still healthy when you decide to have a child, but the frozen eggs are there if you need them and they are the same biological age as you were when you froze them. They are a back-up plan. If you can get pregnant the natural way at thirty-nine, that's great. If you do end up struggling, you'll be glad you had these eggs on ice.

"I am not one of those women who put off having children because of my work, even though I have had a successful career. I just haven't found the right relationship," said Kate, a thirty-four-year-old IT consultant who attended our Denver event with Gretchen. "For me, having a fertility diagnostic assessment done and freezing my eggs is more about peace of mind."

❀ *FERTILITY ASSESSMENT*

Kate and Gretchen each decided to wait a bit before moving forward to freeze their eggs. Although, they opted to sign up for fertility assessments offered by our clinic as a first proactive step. I recommend this even for women still in their twenties because it provides them with the information needed to make critical decisions about their ability to have children. For about seven hundred dollars, we offer a consultation with a physician, an ultrasound, various blood tests, and final consultation to review the results. We determine the condition of your eggs and whether you have any potential fertility challenges that need to be addressed or taken into consideration. We advise women in their mid-thirties to have an assessment to determine the condition of their eggs. This will tell them whether they are healthy enough to be frozen or if they need to take other measures to protect their fertility. Some find that their eggs are in great shape, while others have eggs in a less desirable condition.

A fertility assessment is merely a starting point for making decisions on your timetable for having children. If your eggs are healthy, you have the option of freezing them until you are ready to have a baby. If your eggs are not healthy, or there are other issues, you can begin addressing them early enough so that when the time comes, you are in a better position to have a child. I recommend this for all women, even those still in their twenties, because we complete very thorough tests that can detect most potential challenges.

❀ *THE COSTS*

A major concern for Gretchen and Kate—and many others—when weighing the pros and cons of egg freezing is the cost.

There is the initial cost of retrieving, freezing, and storing the eggs, and then the added expense of in vitro fertilization. Major corporations, including Apple and Facebook, have recognized the importance of the egg freezing option. They pay up to twenty thousand dollars for female employees to freeze their eggs as part of their benefits package.

For now, those farsighted companies are the exception. Most insurance companies do not cover it unless it is necessary to protect fertility because of cancer treatments, but that may eventually change as more companies see this as a beneficial option for their employees.

Costs currently range from seven thousand dollars to fifteen thousand dollars per retrieval cycle. Storage fees to keep the eggs frozen usually begin around three hundred dollars to four hundred dollars a year, often with the first year free, but can run up to one thousand dollars at some clinics. When a woman is ready to become pregnant there is the additional cost of thawing the egg, fertilizing it, and then transferring the embryo to the uterus. "It is quite expensive by the time you've gone through all of the procedures," Gretchen said. "If you think of the price as insuring a baby for the future and compare it to something like the cost of a car, it puts it into perspective a bit."

❀ FERTILITY FACTS

We work with many professional women at our clinics around the country. Women are waiting longer and longer to have children. The challenge for us in trying to help them is that nature doesn't follow social trends.

The science of reproductive medicine is evolving rapidly, but it is generally still true that the optimal age for having children is between twenty and thirty years old. Fertility declines a little in the early thirties for most women and then the chance of getting pregnant can decline rapidly in the late thirties and early forties. Most women who freeze their eggs are already in their late thirties, though we encourage women to do it earlier.

The upper age limit for women to freeze eggs is early forties, but only if their eggs are judged to be healthy. It is possible to have healthy eggs at that age, but not typically. It's also possible for a woman in her twenties to have eggs that appear to be much older. This is why it is so important to have a fertility assessment at a young age, so you know where you stand and what your options are.

Each woman is born with all the eggs she will ever have. Most carry around seven million eggs while in the womb, and that number begins declining at birth. Baby girls enter the world with about two million eggs. The remaining eggs begin to deteriorate and are reabsorbed by the body. This occurs slowly at first.

When a girl reaches puberty, she usually has more than three hundred thousand healthy eggs remaining. During each period of ovulation that follows, a single mature egg is released and available for fertilization. The body absorbs the thousand or so mature eggs that are not released in each cycle.

Only about four hundred eggs go through ovulation. The number and quality of eggs declines more rapidly in the ten years prior to menopause, which generally leaves only about a thousand eggs within the ovaries. Between the age of thirty-five and thirty-seven, the typical drop in the number and quality of women's eggs is substantial. By the time a woman reaches her forties, she may not have enough healthy eggs left. The only option for creating a child in those cases may be egg donation.

Many of our older patients have said, "I wish I'd had the egg freezing option when I was younger." That is why we have worked to get the word out in the media, and events like the Egg Freezing Party. We don't want women to take their fertility for granted. This is one of life's most important gifts and one that requires more thought and planning than simply saying, "If it happens, it happens." The problem is that if you wait too long and it doesn't happen, your options can become far more limited unless you have healthy young eggs safely stored and ready to be fertilized.

❁ LARA'S PERSPECTIVE

Knowing that she can still have children even into her forties and fifties is "a priceless gift," said our patient Lara, a forty-year-old PhD physiologist, researcher, and weight loss coach at a university. "Having my eggs frozen lifted this weight off my shoulders and gave me an indescribable feeling of freedom. If you are a career minded woman who is not willing to settle for an unfulfilling relationship just to be married and have kids, freezing your eggs is incredibly liberating."

Lara spent nearly ten years pursuing her master's and doctoral degrees and she had little time for dating. When she reached the age of thirty-five, she realized her options for motherhood were dwindling. "I wasn't in a relationship and I thought I still might want to have kids someday. That put a lot of pressure not only on me, but on any relationships that developed," she said. "Freezing my eggs removed all of that. It was a life-changing moment for me."

"Now I joke that I'm thirty-seven forever because when it comes to fertility, it's not about your age, it's about the age of your eggs," she said.

Lara's story is particularly interesting because it reflects a surprising, but growing trend in which relatives or loved ones "gift" the cost of egg freezing to women. In the past, a typical graduation gift for a young woman might be a car or a trip to Europe. Increasingly, we are hearing of female graduates receiving funds for egg freezing instead.

We've seen several cases in which parents and grandparents have done this because they look forward to grandchildren or great-grandchildren one day. In Lara's case, the benefactor was a close cousin, whose wife had experienced fertility challenges before receiving treatments from our clinic that helped her attain motherhood. He announced the gift for Lara as a celebration for the completion of her doctoral work.

"I am an only child and my cousin is like a brother to me," she explained. "After I received my PhD, he and his wife offered to do this for me if I was interested and I said, 'Absolutely!'"

Her cousin covered about twenty thousand dollars for all of the tests and procedures, and Lara is now paying three hundred dollars a year to store her frozen eggs. "You can't imagine the gratitude I feel for this gift every time I see him," Lara added. "This gives me the freedom to choose when I have children."

Lara saw that over the same time span she was completing her advanced education, many of her friends married and divorced and she has no desire to follow that path. She describes herself as "happy as a clam" to be single and she has no regrets about waiting for the right person.

"There is so much volatility in the relationships of our generation," she said. "I'm not saying it's a good or bad thing, but many of us grew up in split families or with single parents. I would much rather be single and happy than unhappy in a bad marriage."

Lara also advises other women to have their eggs frozen before the age of thirty-five if possible, because the younger you are, the healthier your eggs will be—and also because if your frozen eggs are younger than thirty-five, you can donate those you do not use to women who need them.

❀ THE SCIENCE

As a health professional, Lara was intrigued by the science of the egg freezing process as well as the insurance it provided for her fertility. "Reproductive science is a fantastic field. You can see all of these follicles in the ultrasound of your ovaries and they keep getting bigger and bigger. It is fascinating to watch that happen," she said.

We have been successfully "slow freezing" human embryos (fertilized eggs) for more than thirty years for use in IVF procedures and the trend to use frozen rather than fresh embryos is increasing. Yet, it's only in the last decade or so that we've been able to develop a successful process for freezing human eggs, which were once thought to be too fragile to survive this type of preservation.

Human eggs, comprised of a single cell, contain so much liquid that the process used for freezing embryos didn't work. Ice crystals formed and damaged the eggs. Because of this, success rates for freezing eggs were poor.

In 2006, flash freezing, or vitrification, began to be utilized routinely for egg freezing. This method avoids the damage to the cell caused by ice crystal formation and the "chilling" effects seen with slow cooling. With vitrification, eggs are taken from room temperature to -196° C in a fraction of a second with high concentrations of cryoprotectants.

Many thousands of babies have been born from frozen eggs using this process and the birthrates from them are now comparable to those for fresh eggs. One recent study at New York University reported thirteen babies born out of twenty-three cycles with frozen eggs.

Our clinic has completed egg-freezing cycles on more than six hundred women including egg donors. Although some of the eggs are frozen for long-term storage and fertility preservation, more than five hundred fifty egg-thawing cycles have been completed with an oocyte survival rate of greater than 91 percent. The pregnancy rate following egg thawing, fertilization, and embryo transfer is 64 percent with a live birth rate per embryo transfer of 55 percent as of August 2018.

Based on our findings, it is likely that women can expect outcomes from fertility preservation similar to those using fresh eggs with IVF. Given our current data, if a thirty-four-year-old patient undergoes egg freezing, she can expect a pregnancy rate typical for women of her age when she thaws, inseminates, and transfers resulting embryos. In rare circumstances, it is possible that oocytes may not survive the vitrification and warming process.

❋ THE PROCEDURES

Reproductive technology and science is advancing rapidly. Now, you can take fertility drugs for ten days and trick your system into making twelve to fifteen eggs, or more, at once.

You do have to take injections—shots—for ten days, so be aware of that. These hormone injections stimulate the ovaries to produce more eggs than usual. During this process, you are required to come in to the clinic so we can monitor your ovaries with vaginal ultrasounds.

Most women feel some bloating and moodiness for a short time due to the hormones. Only about 5 percent of women, most of them younger, respond so strongly to them that their ovaries become swollen and painful. This is called ovarian hyperstimulation syndrome (OHSS) and it can cause nausea, vomiting, and abdominal pain, but it can be managed to avoid serious complications.

After going through the process, Lara said she would advise those who decide to have their eggs frozen to keep an open mind, ask questions so that you understand the process, follow the protocol, and commit to doing it correctly.

"It's not easy because there is a lot of timing involved and you have to give yourself injections. It is in intricate procedure with a very regimented protocol, but it is not difficult if you commit to it," she said. "There were injections to the stomach, but I didn't find them bad. The entire procedure was shockingly easier than I thought it would be. I had a relatively smooth process all in all," she said.

After your round of hormone injections, we retrieve the eggs while you are sedated. The procedure takes only about ten minutes and you won't feel a thing. The doctor uses the ultrasound to guide a needle through the vagina to the ovarian follicles containing eggs. The eggs are removed through a suction device attached to the needle.

"My eggs came out beautifully. We got thirteen healthy, mature eggs and six eggs that matured overnight so I had nineteen in all," said Lara. The only side effects she incurred were a couple of emotional days due to the hormones and some bloating. "Even after the surgery to remove my eggs, I had no soreness, and it didn't seem like I'd even had surgery," she said.

"Being a scientist, I was open-minded and excited to do it, and because I'm not trying to get pregnant right away, it was stress free," she added. "You definitely feel the hormones and feel something a little funky, but you are back to normal in a few days."

After retrieving the eggs, we then drain the fluid from them. Those that are mature are put into cryo-protectants, which work like antifreeze for your car. Your eggs are full of water so they have to be dehydrated; otherwise the water freezes and ruptures the cell membrane. The egg is shrunken like a raisin through the dehydration process, then put in liquid nitrogen and frozen.

We know for certain that the eggs can remain frozen for at least ten years and probably longer. So, when you are ready to have a baby and can't do it the old-fashioned way, you can have your eggs fertilized by your partner's sperm. The eggs will be the same age you were when they were frozen.

The average age of patients who freeze their eggs in our clinic is now around 36 years, but we have done the procedure for women who are forty. Some of them may decide not to have children until they are forty-five or older, but if they do, we require them to see a high-risk pregnancy specialist to determine if their bodies are strong enough to handle a pregnancy. Half of all pregnant women over forty-five experience complications such as gestational diabetes, premature labor, and preeclampsia (hypertension). For that reason, we take extra precautions for women who want to have children after age forty-five.

Our retrieval numbers at CCRM show that egg freezing has increased tenfold between 2010 and 2015. While this process is growing in popularity around the country, egg freezing is still considered "experimental" by the American Society of Reproductive Medicine (ASRM). This is because they are still collecting long-term data on outcomes including pregnancies from thawed eggs. The ASRM does not discourage egg freezing, but prefers to remain cautious until more long-term studies are completed.

❀ A SPECIAL BLESSING FOR CANCER PATIENTS

As I noted earlier, the original group of candidates for egg freezing were those young women facing treatment after being diagnosed with cancer. This is a great service for them, a blessing at a time when they may feel great stress. Since most cancer treatments involve radiation, chemotherapy, or a combination of both, ovarian function is often compromised, rendering the patient infertile. If the patient wants to have children after cancer treatment, freezing her unfertilized eggs will provide the opportunity to do that.

We recommend offering the egg freezing option as soon as possible after a cancer diagnosis. This gives more time to proceed with egg collection and freezing before treatments begin. In addition, patients with nonmalignant diseases, such as lupus and rheumatoid arthritis, may be on chronic medications that are detrimental to oocyte function, so they too could benefit from egg freezing.

The tests we do during the fertility assessment are very thorough and we have, on occasion, discovered for the first time that a patient has cancer. This happened with our patient Giuliana Rancic, the television personality who has written books and given interviews about her experiences. Her case is compelling because during the assessment we found that Giuliana had breast cancer. Since she was married and eager to have children, Giuliana opted to create embryos rather than freezing her eggs.

Cancer is never a welcome diagnosis, but today most cancers can be treated successfully when discovered early, as was the case with Giuliana. There is always concern too that pregnancy can cause a surge of growth in an undiagnosed cancer, which is why we are so careful to screen for it in our fertility assessment. The positive side of this is that if we are aware of the cancer, we can retrieve and preserve eggs so that the patient can still have children after treatment.

Our clinic is fortunate to have Doctor Laxmi Kondapalli on our staff. She is at the forefront of an emerging field of medicine, called oncofertility, which blends the disciplines of oncology (cancer treatment) and reproductive medicine. She is an expert in both cancer and fertility. Laxmi is known around the world for her research and expertise in helping women with cancer protect their fertility.

Her expertise is valuable because often women given a cancer diagnosis have to act quickly to freeze their eggs before beginning radiation or chemotherapy treatments. Basically, they have only one shot to freeze their eggs, so it's critical to do it right. She will tell you more about her work in Chapter Eight. For now, I just want to let you know that egg freezing offers women diagnosed with cancer an opportunity to have children after going through radiation or chemotherapy, which would otherwise not be possible.

❂ *FROM SINGLE WOMAN TO A MOTHER OF FOUR*

I thought you'd like to hear from someone who froze her eggs as a single woman and then married, thawed her eggs and went through the IVF process to have children. At the age of thirty-four, Alexandra was a high-earning, frequent flying, sales consultant for a major tech company. She loved her job, yet her personal life was in the doldrums. She'd just ended a relationship and she couldn't shake the fact that her heavy workload and constant travels were sabotaging her dreams of marrying and having children one day. Then, on yet another business flight, she read an article about the benefits of egg freezing for professional women who, for individual reasons, were facing late entry into motherhood.

"I was one of those girls who'd been a bridesmaid in many weddings. Watching my girlfriends marry and having babies, I felt like I was really behind. I wanted to find a husband and have kids, but I didn't know when, or even if, that was going to happen, so I decided to explore this option," she said.

Her explorations didn't take her far. Alexandra discovered that she lived practically in the backyard of our Lone Tree clinic in suburban Denver. She set up an appointment and came in. Her initial enthusiasm was dampened somewhat as she sat in our lobby waiting area. In that moment, the reality of her situation hit her hard, Alexandra recalled.

"I'd never thought I'd have to do something like that to make sure I'd have a family one day," she said. "It was a very depressing feeling at first. Then, I met Doctor Rob Gustofson and he was so funny and positive and he made me actually feel normal."

Rob, who is a caring physician, listened as Alexandra shared her thoughts and feelings. He noted that her concerns about her diminishing level of fertility were tied to the fact that her married older sister had faced challenges, but eventually was able to get pregnant and have three sons.

Based on her sister's experience, Alexandra knew we did very thorough examinations and tests prior to the egg freezing process. As it turned out, our tests revealed that she did have issues that, if untreated, would have made it difficult for her to become pregnant.

Alexandra had an ovarian cyst as well as endometriosis. Before we could stimulate her ovaries to do the egg retrieval, we addressed those challenges. We found that the cyst was benign, thankfully, and after six months of treatment and monitoring, we gave her the all clear.

Alexandra was glad to finally go through the entire process of hormonal stimulation and egg retrieval, but the physical and emotional effects were tough for her.

"My stomach was distended and I was very moody," she recalled. "It was like PMS. I was edgy and had a diminished capacity to deal with things reasonably."

She handled the hormonal craziness by staying focused on her goal to preserve her fertility. Knowing that she was taking charge of her future helped soothe her and gave her peace of mind. By the time she went through the procedure to retrieve and freeze her eggs, Alexandra was thirty-seven years old. The good news is that during this period, she met and began dating her future husband, Joe.

"We met on eHarmony, the online dating service, and I knew on the second date that I could marry him," she said.

Some of our patients who meet men during and after the egg freezing process have remarked that they often wonder if telling them about their frozen eggs is a good idea, and if so, the women wonder about the timing of that revelation. They don't want men to rule them out because they might think they are too old to have children, but they don't want to scare them off by seeming over-eager either.

I'm afraid that is one question I can't answer. I'll defer to Alexandra. She had no problem telling her future husband about her fertility insurance policy while they were still dating, but only after they'd begun talking about their future together.

"Our conversation on my frozen eggs came at a time when we were already talking about marriage and children so it was very organic," she said. "I told him that having children of my own was very important, so I'd taken charge of my fertility by taking advantage of the science as an insurance policy. I owned it!"

Six months after their first date, Alexandra and Joe were married. He is a couple of years older than her, so they were eager to start a family. She was grateful that we'd already taken care of the ovarian cyst and the endometriosis because those two issues could have delayed her pregnancy for at least a year.

Fifteen of her eggs were fertilized and six proved viable. We transferred one embryo, and it split, resulting in identical twins. Nine months later, Alexandra delivered two boys. By then, she'd jumped off the fast track at work. She took a position two levels down that eliminated the constant travel and allowed for her to work from home with reduced hours. She enjoyed motherhood so much that Alexandra returned to us a couple years later for another successful round of IVF and the result was yet another set of twin boys for her and her husband.

In our most recent phone conversation she was on the road again, but instead of traveling on a business flight, she was driving "a monster Yukon SUV" with her four boys in two rows of car seats behind her. She mentioned something about an ice cream cone battle raging in those back rows. I told her that while we are excellent at helping women have children, that was out of our jurisdiction.

CHAPTER SEVEN

THE EGG DONOR OPTION

BY DOCTOR ROB GUSTOFSON

Peter and Denise tried for almost two years to start their family without success. Denise was twenty-eight when they came to our clinic in Denver. They'd met in college and been married for four years.

The healthy couple became frustrated when they were unable to conceive a child. Our clinic had them go through the standard battery of initial tests and we found that Denise's eggs were poor in quality and quantity.

"Our fertility doctor told me that I basically had the eggs of a forty-year-old woman," she recalled. "I felt blindsided."

Denise and Peter both struggled emotionally with the disappointments, setbacks, and frustrations that so many individuals and couples face when dealing with fertility challenges. In Chapter Eleven, she and Peter share detailed insights into that aspect of their journey and they note the benefits they derived from participating in support groups and organizations for those in similar situations. They also wanted to share their thoughts on the egg donor option based on their experiences.

The CCRM team initially tried to help Denise use her own eggs. She went on a gluten free diet, vitamin supplements, and weekly acupuncture treatments in an effort to improve the quality of her eggs. During this process, she began experiencing severe pain on her right side. When she came to us, we found that she had a large cyst on her right ovary. We removed the cyst along with that ovary and the fallopian tube. After she'd had a few months of recovery, we found that her FSH levels were twice what they should have been. The cyst had been the issue. Her body was working double time to produce eggs.

"There were so many things to try and with most people there is an easy fix, but when we found my FSH was so high, it was a turning point. We missed our second attempt at intrauterine insemination and I thought this was a sign," said Denise, who struggled with feelings of guilt and shame throughout much of her journey with infertility.

There was also concern that the condition of Denise's eggs presented a greater risk of birth defects or other genetic issues even if she did get pregnant.

"Red flags started going up and we had to ask why we were focusing on using my damaged eggs," she said.

Peter and Denise decided to consider options beyond using her eggs, which included adopting a child, using donor eggs, or choosing to be child free.

"Our doctor said there was a low probability of in vitro fertilization working for us, so we took a hard look at all the options. What mattered most is that we really wanted to have a healthy family and would get there any way we could," said Peter.

They talked with two adoption agencies, one local and another international, to study that option. They also researched what was involved in using a donor's eggs. There was so much to consider and they were still dealing with the emotional pain of infertility. In the summer of 2011, they decided to hit pause and take a two-week vacation from the stress.

"We drove to California and talked the whole way," said Denise. "It was nice to be in an environment with no distractions without daily doctor visits."

During the trip, Peter and Denise decided on a course of action. Although the idea of using a donor's egg was still a foreign concept, Denise was still able to carry and experience that rite of passage to motherhood.

"We just kept getting more and more excited about using a donor's egg and it was a huge relief," Denise said. "They'd told us that there was less than a 3 percent chance of having a child with my eggs, but with a donor's eggs there was an 80 percent chance of a successful pregnancy."

Peter recalls that once they'd made the decision to use a donor's eggs, "Our whole outlook changed. We went from having almost no chances of having a pregnancy to a probable one!"

Once the focus was off of using Denise's eggs, she felt relieved to be moving on. "I still had the ability to carry and didn't want to miss out on that opportunity," she said.

❁ *THE LAST BEST HOPE*

When we see patients who need donor eggs, most of the time they have accepted this option only because it is their last best hope. It's rarely a first choice for them, but using an egg donor is often the last best choice. Doctor Schoolcraft often tells women trying to make a decision that if they compare using a donor egg to using their own, it doesn't make sense. But if a patient's eggs are not viable, the options come down to giving up, adopting, or finding an egg donor.

Another consideration is the substantial financial cost involved. The approximate cost of donor egg IVF in our program with donor work-up, medications, and an anonymous egg donor cycle is over thirty-five thousand dollars. The entire process, from evaluation to the actual transfer, can take from two to nine months to complete.

With an egg donor, the husband's sperm (or in some cases, donor sperm) is used to fertilize the egg so that the child will carry the father's DNA, and if you find a donor who matches up well with the patient, all the better. In fact, Denise and Peter are among many couples who've used egg donors and discovered that their children born from the donated eggs are surprisingly similar to them in many ways. The fact is that half of the child's forty-six chromosomes come from the father and the other half from the egg donor.

However, when those chromosomes are translated into genes inside the womb of the patient-mother, magical things can happen. The translation of the chromosomes to genes and subsequently proteins are influenced by the mother, her hormones, and her nutritional status.

Scientists believe factors including diet, environmental pollutants, and even stress can impact how the fetus develops. Our genes are templates to produce proteins and they are influenced by the environment in the womb. So, the mother carrying the child has a lot of influence in terms of what genes are transcribed and what proteins are produced. This is a science called *epigenetics*. Many mothers who use donated eggs find that exciting because it makes them feel such an important part of the baby's development.

It's important to women on an emotional level to feel that the child is theirs and that can be a major issue to consider when deciding whether or not to use a donated egg.

Pragmatically, however, a donor egg is often seen as the best choice if the other options are giving up or adopting.

Now, here is the kicker: When our patients come back after having the embryo transfer and becoming pregnant, they are often "over the moon" with excitement and joy. Typically, they say things like, "I realize that without my body and my uterus, this baby would not be born. I truly feel that I am the baby's mother, and if I'd known that I would feel like this, the decision to use a donor egg would have been much easier to make." There are other factors to consider, including very good success rates for the egg donor option. Success rates from donated eggs vary from clinic to clinic. In 2016, our clinic completed two hundred ninety-nine egg donor transfer cycles and one hundred ninety-four resulted in pregnancies for a success rate of 73 percent.

That is another reason why donor eggs (oocytes) have become a very viable option for women who may have no other chance to become pregnant. This treatment can lead to successful pregnancies for women with any of the following fertility challenges:

- Premature ovarian failure or early menopause
- Eggs of poor quality judged by decreased egg counts or a prior poor quality IVF cycle
- Older reproductive age group (under age fifty)
- Chromosomal translocations or genetic diseases that they wish to avoid passing on to their offspring

The decision to use an egg donor can be very easy for some when faced with premature ovarian failure or very difficult for those who have poor quality eggs. Most women would prefer to use their own eggs, naturally. After exhausting all options with their own eggs through either IVF, alternative therapies like acupuncture, or considerable time trying, some decide to use donor eggs as their only hope for having children.

We find that once a woman or couple accepts donor egg as an option, they embrace the idea and often move forward knowing that they have tried everything in their power to conceive on their own.

Compared to the process of using donor sperm, the process for obtaining donor eggs is more complex. Eggs can be used fresh, or vitrified eggs can be thawed and fertilized. Donor sperm can be frozen for decades and obtained from sperm banks and now eggs can be too.

Until recently, egg freezing has not been successful and thus, not a viable option. In coming years, we expect this to change due to the increasing success of rapid egg freezing methods known as "egg vitrification." With this method, donor egg banks will likely begin to emerge and greatly simplify the process just as donor sperm banks have done for others.

❀ *FINDING A DONOR*

Currently there are four options for selecting egg donors. They include:

I. Option 1: Fertility Clinic Donor List

Many fertility clinics develop their own anonymous donor database. In most cases, you pick a donor based on criteria such as physical appearance, ethnic background, education level, and genetic family histories that run three generations back and include information on incidences of cancer, diabetes, and other illnesses. At our clinic, we recruit a host of donors from surrounding communities.

Our donors are often college students, young professionals, or young mothers. The benefits of using your clinic's donor program usually include less expense because their compensation is fixed without travel or agency fees. We offer photos of the donor as a child but not as adults because we want to protect their identities and most of them live in our area.

If you decide to use a friend or family member or a donor provided by an agency, you select the donor and your clinic performs medical screening to ensure safety for you and your donor.

When one of our patients uses a donor from our clinic's pool, the patient chooses her from our online portal after receiving her log-in. The patient chooses the donor according to their personal preferences. Most women or couples look for a physical match with their donor. Others require different qualities like musical ability, intelligence, or athleticism.

Specific ethnic groups can have a tougher time finding a donor. We had a Nigerian couple who looked for more than a year to find a donor who shared their ancestry. Asian couples can also be very specific about wanting to match up with their own particular group, whether it is Korean, Vietnamese, Chinese, or Taiwanese.

Many patients looking for egg donors find themselves pondering the age-old question: Is it nature or nurture that creates the child? Is an egg donor with musical ability necessarily going to help you have a musical child? Does a donor's high intelligence guarantee your child will ace organic chemistry in college? Or does family environment have more to do with a child developing certain talents and abilities?

We can't give you those answers. Science has struggled with the nature or nurture question for generations. If you believe nature, then you will often seek a donor with the qualities that you want exemplified. If you feel the environment decides their fate, choose a good physical match and raise your child in the setting that suits you best.

Our egg donor program is unique in that all donors on our list must be screened and ready for donation as soon as you accept them. This saves the heartache that may occur when the "perfect" woman doesn't pass her screening and is then not eligible to be your donor.

II. Option 2: Donor Agencies

These are companies that help people find egg donors. They are matchmakers of sorts but they are not medical practitioners and don't do the procedures.

If you find a donor working with an agency, you then have to take that donor to a clinic to screen and have eggs harvested and embryos created. An egg donor agency recruits and retains donors from anywhere in the United States. Various agencies focus on recruiting different types of donors. For example, an agency in New York specifically recruits only donors of Asian descent, while another agency in California seeks intellectually gifted donors with high IQ scores, high grade point averages, and top test scores.

Essentially, you have a much larger pool to draw from by using an agency to find a donor and this is often touted as one of the best benefits of this option. Most agencies are reputable and recruit high quality donors. You should check them out by asking for references and talking to people who have used them.

The benefits of using an agency donor is the wide selection offered. The downside is that after the patient has chosen a donor, that donor still needs to pass our clinic's work-up requirement, which could mean time and money lost if she does not qualify. Often, agencies offer photos of the donors as adults and a list of their characteristics without identifying where they live to maintain anonymity. The downside is cost—the agencies charge a fee for finding the donor. Some donors may command a higher fee based on her characteristics and occupation. There also may be travel fees to pay, depending on your egg donor's location and its distance from your clinic.

Each agency has its own fee scale. We advise you to carefully check out any agency you choose to deal with before you sign an agreement with them. Keep in mind that some fees, regardless of whether you use their donor or not, are non-refundable.

III. Option 3: A Known Donor

This is a fairly common option, and it is sometimes known as "the Friends & Family" plan. It involves recruiting a relative or friend who volunteers to be an egg donor on your behalf. The screening is still the same for your clinic, which will check the donor for infectious diseases and quality of eggs. If you use a known donor, you will need to have an attorney draw up a contract that spells out the agreement and calls for the donor of the egg to relinquish all claims and rights on the child.

The friends and family option has many excellent benefits. This egg donor is well known to you and even may be genetically linked. The process tends to be less expensive because you don't usually have to pay them. The downside of a friend or family member as donor often occurs during or after pregnancy when relationship issues surface.

To save you both grief, you and your known egg donor should have heart to heart discussions with a counselor early on about what sort of relationship they'll want with you and the child after birth. We require this for every patient using any form of third party reproduction. For example, if your sister is your egg donor, will she be satisfied to merely serve as the child's aunt or will she feel more like a mother to the child? If she feels more motherly, this could lead to conflicts with you as you raise the child.

Friends and family can be a wonderful choice for egg donors. The caveat is that you should have very frank, open, and honest conversations with them prior to the first step. This conversation may be difficult, but it is necessary to avoid conflicts and misunderstandings later. You should work through what you want and what they want, as well as what both sides expect.

IV. Option 4: Egg Banks

This is a relatively new option, but one that could be expected to develop in a transactional world. Many are independent of clinics and will work with any IVF clinic. They may have sixty or more donors available that they feature on one website.

The egg bank harvests eggs from donors and freeze them. Often, the participating clinics may be in different cities so the donor eggs a client selects may be shipped to your clinic where they would be thawed and fertilized to create an embryo.

✿ ONCE A DONOR IS CHOSEN

Denise and Peter chose their egg donor from profiles in the CCRM donor pool.

"We narrowed it down from about sixty to twelve and within a week we were down to two donors, but we kept going back to one profile because she seemed like someone I could be friends with," said Denise. "We had her whole family history—twenty-seven pages of it—going all the way back to her grandparents."

One of the things about this donor that ultimately led to Peter and Denise choosing her was the fact that she'd known other women who'd had difficulty getting pregnant and she wanted to help others in that situation. That sealed the deal and once they chose this donor "it was full speed ahead," they said.

"It had been a three-year roller coaster ride, so we were really ready to become parents," said Peter.

There was one more option that they had to consider. Did they want just one child, or more? They had been a couple without children for a long time. After encountering a pair of very active two-year-old twins at a support meeting for couples using donor eggs, Peter asked Denise if she was sure she wanted twins. Denise responded that she'd rather go for two children than none at all, but ultimately it is based upon the physician and ASRM recommendations, patient history, and quality of the embryos. After review, the physician makes the recommendation of the number to transfer.

As things turned out, Peter and Denise did end up with twins.

"When the nurse said, 'You are very pregnant!' I was overwhelmed because I'd never had a positive pregnancy test," Denise recalls. "And then I thought just how many are in there?"

Both Denise and Peter had fought infertility so long that they admitted to feeling a little sad about leaving their infertility support group friends to join a support group for couples who were about to have twins. But once their twins were born, they happily acknowledge that their lives had changed forever.

"When your children are in your arms, your perspective changes in every way," Denise says.

Peter is only half-joking when he says that he and Denise are now living with a daily science experiment. Although their children, Ophelia and Olive, were conceived with a donor's eggs, Denise carried them in her womb and they suspect there was some intermingling of DNA during pregnancy.

"Our donor's profiles said she was very athletic, had straight teeth and perfect eyesight, which are not traits I share," Denise says. "So I thought those were a positive. Now, looking at our daughters, people say one is very similar to me and one is like Peter. They have sandy blonde straight hair and our donor had brown wavy hair. The donor had brown eyes and they have blue eyes, like Peter. So, it's very interesting to see their characteristics."

The couple still has three frozen embryos and while they have not ruled out having more children, they also have given some consideration to donating embryos to other couples dealing with infertility. "If it weren't for our donor, our family would not be what it is today," said Denise.

"Right now, we have twin toddlers and it's enough for us, but we go back and forth on what to do with the embryos," Denise said. "We've thought that maybe we ought to give another family a chance. So, we are taking our time on this decision."

Another issue that they are still weighing is one pondered by many parents who use an egg donor; the issue of whether or not to one day tell the child and if so, when to share it.

"We've discussed this in our support group, with our counselor and other parents. Some of our friends are open about using an egg donor with their children and we are among the most open couples," she says. "Still, some people aren't open to sharing. My feeling is that we want our children to trust us as they grow up and we wouldn't want them to ever wonder why we hid something as important as this from them."

Denise and Peter said they would rather be open than worry that someday their children would take the popular DNA tests or do a genealogy history. They say these days the world is much more open regarding fertility treatments and infertility so there is much less concern about children being ashamed of their origins if an egg donor was involved. The couple said they wouldn't be surprised if one day their twins try to track down the egg donor.

Our counselors generally advise openness to patients at CCRM. We've found that if children are told early on, starting around the age of two, that their mothers struggled to get pregnant and wanted them so badly they asked for anothr woman's eggs, the kids accept it and grow up comfortable with it.

Companies like 23&Me, the DNA testing service, are increasingly popular with people who want to sequence their DNA and learn about their origins and connect to relatives. A child born with the help of an egg donor might be shocked to learn he or she isn't related to a parent, so again, we think it is best to tell them at an age when they are mature enough to process the information and accept it.

❀ A DONOR'S STORY

Martay was in her early twenties, single, and in a long-term relationship when she served as an egg donor on two occasions. The first was with our clinic, the second was with another clinic as she had taken a position at CCRM and was therefore not permitted to be an egg donor for us, a policy meant to avoid any conflict of interest issues.

This adventurous and compassionate young woman decided to become an egg donor after seeing many advertisements soliciting egg donors by clinics in our area. She grew up in a big, loving family that "took in strays" of all kinds so she has a natural, giving and kind spirit.

"I didn't know anyone who struggled with infertility, but I had empathy for any woman who faced it and I thought I was at a good stage and time in my life to do this for someone else," Martay said. "My boyfriend and my family were fine with it, I am very healthy, so I went for it."

We thought it would be helpful to share the perspective of an egg donor and Martay agreed to offer her experiences, both good and bad. Being an egg donor can be very rewarding, yet it is also time-consuming, invasive, and requires many office visits and injections as well as blood draws. The donor must be totally committed to the process because of all that is required of her.

❀ *A BIG COMMITMENT*

Egg donation often takes time away from the donor's everyday life or work. They are compensated for their time and this may represent a motivation for some donors. In our program, first time donors are compensated up to eight thousand dollars and repeat donors receive additional incentive. In comparison, the average sperm donor receives approximately fifty to one hundred dollars per sperm collection.

How can you adequately compensate someone for giving you their life-giving eggs? The answer is that you cannot. Their compensation is solely for their time and effort expended during the cycle. Recipients cannot buy eggs—they can only compensate the donors for their time. According to the American Society for Reproductive Medicine (ASRM), "monetary compensation of the donor should reflect the time, inconvenience, and physical and emotional demands and risks associated with oocyte donation and should be at a level that minimizes the possibility of undue inducement of donors and the suggestion that compensation is for the oocytes themselves."

Ads in an early 1990s newspaper offered fifty thousand dollars for the perfect donor—clearly an excessive amount. This excess is strongly frowned upon by most, if not all, legitimate infertility centers and by the ASRM. Egg donors typically spend an average of forty to fifty hours total in the standard procedures. The time spent includes driving to and from the clinic, having various medical evaluations, injections, procedures, and time off of work. Again, it is a significant investment in time, emotion, and physical invasiveness.

● *THE SCREENING PROCESS*

Regardless of how you select your donor, each must complete very rigorous screening in our program, and most other agencies and clinics. On average, we accept only one out of ten donors we screen. We like to say that at our clinic, it is more difficult to be an egg donor than to be accepted to most colleges!

For a friend or family member, the acceptance criteria and allowable age range are less restricted. We look for friends and family donors who are between nineteen and forty years old and preferably less than thirty-five years old. For an anonymous donor, the recommended range is between nineteen and thirty-three years old.

The screening for anonymous donors at CCRM, and most infertility clinics, requires the following:

- A medical history and physical exam with a physician to assess their general well-being

- Testing for infectious diseases in accordance with the U.S. Food and Drug Administration guidelines. This ever-growing list includes HIV, hepatitis B and C, syphilis, gonorrhea, Chlamydia, cytomegalovirus, West Nile Virus, and trypanasoma

- Illicit drug screening

- Chromosomal karyotype

- Screening for inheritable diseases such as cystic fibrosis, thalassemia, and spinal muscular atrophy based on ethnic ancestry

- Review of two generations of family health history by a genetic counselor and physician to evaluate for heritable diseases

- Evaluation of egg quality and quantity

- Psychological assessment by personality inventory questionnaire and interview with our staff psychologist.

In her case, Martay lived just down the street from CCRM, so she decided to apply to be on our egg donor list because of the convenience. She was surprised at how much work went into the screening process. The first step was to fill out an online questionnaire that is designed to screen out anyone who has used illegal drugs, been a smoker or heavy drinker, and other issues.

Once she passed the initial screening, Martay was asked to fill out an even more challenging twenty-page questionnaire that goes into more depth. Still, as an athlete with a competitive streak, she found herself wanting to make the cut as she answered questions about her medical history, physical condition, her childhood family life, history of cancer and other chronic illnesses in the family, hobbies, and life goals. We also ask for childhood photos between the ages of two to seven so those looking for an egg donor can see if there is a physical match to what they looked like as children.

After she completed that extensive process and passed the initial review, Martay then came in for a personal interview and a one-day physical and medical workup, which is very thorough and similar to what all of our patients go through. The next step was a meeting with our staff in which Martay was given clearance as an egg donor and asked if she was ready to commit to doing it.

Once she committed, Martay was added to our list of egg donors and we all waited for someone to choose her as a donor. We assured her that we screened the recipients as carefully as we screen the donors. We want to be sure that there is no other potential cause that may decrease the chances of success other than poor egg quality. The evaluation for potential recipients of donor eggs includes a thorough assessment of every client's uterus—a key factor in the donor egg process.

The uterus is evaluated by ultrasound for evidence of fibroids, polyps, or other structural abnormality. Blood flow to the uterus is reviewed by Uterine Artery Doppler—a very specialized ultrasound that determines how well blood is flowing. An office hysteroscopy is performed to evaluate the inside of the uterine cavity and to determine if there is any scarring or abnormality present.

The final test is a functional assessment of the uterus—the woman uses estrogen to build the lining of the uterus similar to what would happen when the real cycle begins. This is called a mock cycle and ensures that the uterus will respond properly during the procedure.

The male partner will also put in his share of time at the clinic. It is critical to evaluate his sperm or that of the sperm donor in the case of single women or lesbian couples. Both the male and female will be tested for transmittable infections such as HIV, hepatitis, and syphilis.

All parties involved will also be given psychological assessments because donor eggs and pregnancy create major life changes in both wondrous and stressful ways. The psychological evaluation is performed by a counselor trained in infertility issues. The patients are provided helpful guidance for dealing with this process.

❋ GETTING SYNCHRONIZED

Once both the donor and the recipient have completed the screening process, the magic begins. This is an exciting time because the possibility of pregnancy for the egg recipient is just a mere few weeks away! In order to coordinate the women's cycles, birth control pills are often used by the donor and the recipient woman. Yes, it's a rather counterintuitive deal to be using birth control pills since you are trying to get pregnant and not prevent it. Strange as it may seem, the pill is crucial because it helps us align the menstrual cycles of both women effectively.

Once the process is underway, the donor begins injecting medications to stimulate the ovaries. This is done under close monitoring at the clinic. Simultaneously, the recipient starts the estrogen again to build the uterine lining.

Several weeks later, we call the donor in to begin the process of giving her fertility drugs to hyper-stimulate her production of eggs. This process usually lasts from seven to twelve days and requires twice daily injections of the fertility drugs.

Martay's first experience was typical for an egg donor. She doesn't hide her feelings when describing the stimulation cycle in which the donor has to give herself injections each morning and night. "It's icky. You don't feel good. It's not glamorous to go through. I maintained a normal lifestyle throughout the whole cycle, but you don't feel good afterwards. There is nothing about being an egg donor that makes it any easier than if you were doing it for yourself to do IVF. You don't want to mess up the process or leave your meds out of the refrigerator because someone is counting on you to produce eggs for them. I felt a responsibility to do everything right."

The average egg donor has to do the fertility drug injections for nine days, which means keeping to the schedule, limited physical activity, no intercourse because you are hyper-fertile, staying out of the sun, no alcoholic beverages or caffeine, and limited social activity because you have to stay on schedule.

After nine to fourteen days of stimulation, the donor undergoes her egg retrieval—an outpatient surgical procedure to "harvest" the eggs. Her eggs are then fertilized in the laboratory with sperm, either a partner's sperm or donor sperm, and then grown in a culture for three or five days. The patient can also use frozen donor eggs to be fertilized and grown out in the same manner.

Once her eggs were ready for retrieval, Martay came in to the clinic where she was prepped and given anesthesia. "I was a little loopy when I met my doctor, Doctor Gustofson, for the egg retrieval," she says. "I remember getting rolled into surgery and that's it. When I woke up in recovery, they gave me some ginger ale and a cracker and a thank you card with my compensation in it. Later, they do scans and checkups to make sure your body has returned to a normal state."

After surgery, Martay had three or four days of discomfort. "It didn't feel great. I stimulated so well that my ovaries were huge and my comedown time was rougher than most," she recalls. "It was a little like going through withdrawal for two or three days, but at no point did I say it wasn't worth it."

Once fertilization and incubation have completed, the embryos are transferred to the recipient's uterus. The number of embryos transferred is usually one or two depending on the donor's age and success rate. Once transferred, the long wait begins. The nine- to eleven-day wait can be the most grueling of all, so we recommend that you plan some activities to take your mind off what's going on in your body, if that is possible.

❁ POST-PROCEDURE

Often, there are extra embryos available after transfer to the uterus. These may be frozen for use later on; either if a pregnancy doesn't occur or if more children are desired. This frozen embryo option gives patients the chance to have genetically similar children without repeating the entire process.

Although this occurs frequently, frozen embryos are not a guarantee, as the goal is to achieve a pregnancy the first time with the donor. Extra embryos then are a bonus. Sometimes, an egg donor is willing to donate again to a couple who previously used her eggs, so that the children will be true genetic siblings, or if using a different sperm donor, at least genetically linked.

Most women who donate their eggs do so without any complications and they return to their pre-donation health status within a few days. As with any procedure, there are risks involved to the egg donor. Because the egg retrieval is a surgical procedure, there is a minimal amount of risk due to the surgery or anesthesia. The risks include bleeding or infection and may require other surgery to stop any bleeding that may occur.

The donor is given fertility medications and studies have shown no increased risk of cancer or long-term side effects in women having taken fertility medications for up to six donor cycles.

Another risk of donating eggs is ovarian hyperstimulation syndrome (OHSS). This is a rare complication of ovarian stimulation and there are methods to help prevent this from occurring. Often, it can be treated by bed rest and increasing fluids.

Martay later received an anonymous thank you note from the woman who received her donated eggs, but they had no further contact. Martay was called to donate her eggs again at our clinic; however, on the same day the call came, she had accepted a job to work for us, so she couldn't donate with us again.

❀ *A BAD EXPERIENCE*

Martay did later donate eggs for another clinic's program and she ran into some of the pitfalls of egg donation there, which we are relating only as a cautionary tale to others who might want to donate eggs.

Since her first egg donor experience had gone so well, Martay signed up with another clinic to do a "split cycle" in which her eggs would be split between two couples needing donor eggs.

"To make a long story short, it was an awful experience," she says. "I got very ill and I was in the hospital twice."

In this case at the other clinic, Martay was overstimulated, which can happen if precautions are not taken and the donor is not carefully monitored. She suffered from ovarian hyper-stimulation syndrome (OHSS), which is the result of having a large number of growing follicles along with high estradiol levels. This leads to fluid leaking into the abdomen (belly). It usually occurs several days after egg retrieval for donors if medications to prevent it aren't administered properly.

Although it can range widely in severity, OHSS can be very painful and it appears Martay had a fairly severe case. Bloating is common after the ovaries have been stimulated, but it can be so bad that women suffer nausea, vomiting, and feel like they are having trouble breathing because of pressure from the ovaries and fluid on the diaphragm.

Martay gained twenty pounds of fluid and had to sleep on the couch for a week with pillows propping her up.

"I looked seven months pregnant and they kept telling me it was normal, but I learned later that if you gain five pounds of fluid weight you should be brought in for examination," she said. "I went in to the clinic after I'd gained ten pounds and they said I was dehydrated. They called my doctor at that clinic and she told them to give me more fluid and send me home. It was scary because other nurses I knew said I needed to get drained instead. I suffered for two and a half weeks before my body reset itself. It was horrible to be so miserable."

❀ *THE EMOTIONAL SIDE*

Because of that experience with another clinic, Martay decided not to be an egg donor again. Her bad experience highlights the importance of understanding what can go wrong and getting the best possible care at the best possible clinic.

There is also an emotional side to being an egg donor. Martay advised women considering egg donation to understand that they will have feelings surface afterwards. They may find themselves wondering about the child born to the recipient, which is only natural.

"You have to let it go and not obsess about whether the child looks like you," she said. "You have to respect the privacy of the recipient and understand that the child doesn't belong to you in any way. I wondered, of course, and now and then I think how cool it is that someone with my DNA could be out skiing with the parents. I always hope that the child is born into a great family."

CHAPTER EIGHT

FERTILITY PRESERVATION OPTIONS FOR CANCER PATIENTS

BY LAXMI A. KONDAPALLI, MD, MSCE

Television personality Giuliana Rancic generously shared her long and arduous journey to parenthood in the first chapter of this book. That journey included her diagnosis and treatment for breast cancer. Fortunately, she had frozen embryos and the means to use a gestational carrier, which allowed her to have a son after her cancer treatments.

Giuliana is one of many cancer patients who serve as sources of inspiration and hope for others who want to have children. We are grateful that the fertility options and opportunities continue to grow for both women and men diagnosed with cancer.

With cancer survival rates steadily increasing, what was once considered a "terminal illness" is now more of a life challenge event than a death sentence. Yet, it is also true that some of the newer life-saving treatments that increase cancer survival rates can negatively affect fertility. It can do so by causing delays in childbearing or even by compromising a person's ability to have children.

Over the last several decades the emergence of my specialty, *oncofertility*, along with the development of new fertility preservation techniques has made the dream of building a biological family a reality for many cancer survivors.

Still considered an evolving interdisciplinary field, oncofertility exists at the intersection of oncology, (cancer treatment), reproductive endocrinology (a subspecialty of obstetrics and gynecology that treats hormonal issues related to infertility), and fertility care. The goal of oncofertility is to expand reproductive options for young patients with cancer, and it can be a great gift to them.

❀ THE NEED TO ACT QUICKLY AND ASK THE RIGHT QUESTIONS

Each year, approximately 150,000 men and women age forty-five and younger are diagnosed with cancer. Fortunately, the five-year overall survival rate for these patients is approximately 80 percent. An estimated 30–75 percent of males and 40–80 percent of females in this population are rendered infertile as a result of cancer treatments, including chemotherapy, radiation, and surgery.

These statistics make infertility a very real and unfortunate side effect of cancer and cancer treatment. As cancer treatments improve, more people are looking at quality of life after cancer, which includes growing a family. While many young cancer patients can expect long-term survival, we advise all of them to take measures to preserve their ability to have children before they begin cancer treatment. It is critically important that you discuss with your oncologist the possibility of infertility and ask about your fertility preservation options as early in the treatment planning process as possible.

If you or someone you know is diagnosed with cancer and still hoping to have children, I recommend you ask your oncologist these questions as soon as the cancer diagnosis is made:

- Will my cancer or its treatment affect my future fertility?

- What are my fertility preservation options?

- How quickly do I need to start treatment?

- Is it safe for me to carry a pregnancy after my treatment?

- Can you refer me to a fertility specialist?

Timing is critical because most patients cannot and should not delay their cancer treatments, but they also want to do whatever they can to preserve their fertility before radiation or chemotherapy treatments.

Shea wisely took immediate steps to protect her fertility after her cancer diagnosis. She was initially diagnosed with Hodgkin's lymphoma and was found to have a recurrence of her disease eight months after chemotherapy. Her physician recommended a bone marrow transplant, a treatment that would almost certainly render her sterile. She was only twenty-six years old.

After a discussion of her fertility preservation options, Shea chose to freeze her eggs prior to initiating treatment. She was between menstrual cycles, but was given a beneficial drug with a scary name, a *GnRH antagonist*, which caused her period to start within a few days. At the start of this menstrual cycle we stimulated the ovaries with a hormone called follicle stimulating hormone (FSH) to induce her body to grow multiple eggs simultaneously. Eighteen eggs (oocytes) were harvested about ten days later. They were then frozen (vitrified) for her future use. After this brief two weeks of fertility preservation, Shea was ready to start her cancer treatment, and with greater peace of mind. She went into that treatment knowing it would not endanger her ability to have a family one day.

Thanks to the rapidly advancing science across a number of disciplines and medical specialties, young women like Shea have more than a fighting chance of both surviving cancer and having children one day. Let's look at some of the science involved in fertility preservation.

● *THE THREAT TO YOUR FERTILITY*

Women are born with all the eggs they will produce in their lives. So, the goal of oncofertility specialists, like me, is to try to make sure that those battling cancer can use their eggs later on when they are ready to have children. A woman's ovaries contain thousands of follicles, called the *ovarian reserve*. Each follicle contains a single egg.

During each menstrual cycle, a group of immature follicles begins to grow, but only one goes on to fully develop and release its egg during ovulation. The number of follicles in the ovarian reserve naturally declines with age, which is itself a threat to fertility. Cancer treatments like chemotherapy and radiation can destroy the follicles in the ovary, accelerating the natural decline of the ovarian reserve. As a result, some women may face temporary or early menopause after cancer treatments.

When the ovary is subjected to chemotherapy or radiation, the eggs can be damaged or destroyed, but there are ways to minimize this risk. For example, to protect the eggs during radiation therapy that involves the pelvis or abdomen, we can move the ovaries higher in the abdomen to protect them through a minor surgical procedure called *ovarian transposition*.

Another method that has been found to help lessen the impact on eggs, even with the harsh treatments of chemotherapy, is the use of medications that "quiet" the ovaries. Some chemotherapy drugs only affect developing follicles and the eggs inside them. Therefore, a woman may stop her menstrual cycle during cancer treatment, but because she still has immature follicles in her ovarian reserve, she may start having her period again when treatment ends.

Other kinds of chemotherapy, especially those that use alkylating agents that bind to DNA and prevent proper DNA replication, can also damage the immature follicles that are resting in the ovary. Once these follicles are destroyed, women cannot make new ones. Consequently, these therapies can cause early menopause in young women or even delayed puberty in girls. In addition to damaging the follicles, radiation to the abdomen or reproductive organs can damage the uterus, which may make it difficult to carry a healthy pregnancy.

● FREEZING OPTIONS FOR WOMEN

For women cancer patients who have gone through puberty, the established methods for fertility preservation are embryo and egg freezing and banking, or cryopreservation.

Colleen, a twenty-one-year-old, was diagnosed with acute myeloid leukemia. Her treatment outlined by her oncologist required the use of two very strong chemotherapy medications. Her oncologist wanted to start the medications within two weeks of her diagnosis. With the short window of time, she elected to pursue egg freezing and had an initial ultrasound and blood test called AMH suggesting her eggs were plentiful and viable. After 12 days of stimulation medications, she underwent transvaginal egg retrieval with IV sedation, had 27 eggs frozen, and was able to begin chemotherapy the next morning.

❋ THE ADVANTAGES OF FREEZING EGGS INSTEAD OF EMBRYOS

Although the first pregnancy with frozen eggs was reported in 1986, this utilized a technique called slow freezing. It took another twenty years for egg freezing to become successful and mainstream, primarily thanks to the advent of the more rapid freezing method called vitrification. This led to markedly improved success rates with regard to the survival of eggs after freezing and their resultant pregnancy rates.

Indeed, some centers today have success rates with frozen eggs comparable to those of fresh eggs, especially in younger women. Freezing eggs rather than embryos has obvious advantages for young women, particularly those who do not have a partner or a sperm source to create embryos at the time of their cancer diagnosis. The eggs are stored long-term. Later, when the patient has finished her cancer treatment and is ready to have a baby, she can thaw those eggs and fertilize them with the sperm of her partner, or if she were single, donor sperm could be used as well. In addition, freezing eggs rather than embryos avoids the ethical and legal issues surrounding embryo storage and disposal, which is a concern for some patients.

Eggs, however, are more sensitive to the impact of freezing than embryos and thus the success rates are somewhat lower. Eggs are particularly sensitive to the freezing and thawing process for many reasons, including the large volume of water within the cell and the fact that the chromosomes are aligned or tethered to each other by structures called spindle. Rapid freezing, or vitrification, has minimized the negative impact of freezing on eggs and made it a very realistic option for patients.

It was once believed that fertility drugs used to recruit eggs could only be started on the menstrual cycle, thus limiting the start time to once per month. The problem with that was that when patients were diagnosed with cancer they often had to wait up to three and a half weeks to start their treatment. More recently, it has been found that a patient can start the stimulation drugs at any time in the monthly cycle and still get similar results with regard to egg quantity and quality. This has certainly been an improvement of the speed at which cancer patients can be treated.

Lily, a thirty-eight-year-old, was diagnosed with an estrogen receptor positive breast cancer. She also had history of a blood clot. Because of this, she was interested in freezing both eggs and embryos before starting breast cancer treatment. Medications, including FSH to stimulate eggs to grow and a drug called letrozole designed to minimize estrogen levels, were given to her simultaneously.

For Lily, minimizing the estrogen rise is important, given that her cancer is estrogen receptor positive. She was also started on a blood thinner called Lovenox during her stimulation so that any estrogen rise would not lead to another blood clot. Lily was able to freeze a total of eight eggs and eleven embryos, giving her the option to thaw both eggs and embryos to have a baby when the time is right for her to start a family.

Both embryo and egg banking require women to take hormonal injections to stimulate the ovaries to produce multiple mature eggs over the course of one to two weeks. These hormone injections may not be appropriate for women with certain types of cancer, so it's important to ask your doctor if this is an option for you.

The process is closely monitored through blood tests and a series of pelvic ultrasounds. When the follicles reach a certain size, an egg retrieval procedure is performed, often in a doctor's office with minimal anesthesia. The mature eggs collected during the procedure can be frozen individually and stored for future use, or they can be fertilized with a partner's or donor's sperm to form embryos, which also can be frozen and stored to achieve a pregnancy in the future. Once the egg retrieval is completed, a woman can begin her cancer treatment.

❀ *OTHER OPTIONS FOR FERTILITY PRESERVATION*

If a female is too young or is not a candidate for egg or embryo banking, she may consider *ovarian tissue freezing*, a procedure by which the outer layer of the ovary, which contains immature follicles, is surgically removed and then frozen and stored. First, ovarian biopsies are obtained through a minor surgery called laparoscopy where the tissue can be harvested. Later, when the patient has finished her treatment and is cured, the ovarian tissue can be thawed and transplanted back into the patient in or around the ovary where it was previously harvested.

Ovarian tissue freezing is often an option for patients requiring immediate chemotherapy where a delay of even a few weeks is not possible and is the only option for young girls prior to the age of puberty.

Marie, an eighteen-year-old, was diagnosed with Ewing's sarcoma, requiring high doses of drugs that likely would have severely impacted her ability to have children. Because of her young age, the highly aggressive tumor, the strength of the cancer treatments she faced, and the need to start them immediately, Sarah was referred for a fertility preservation consultation. She was extensively counseled about her options including embryo, oocyte cryopreservation, and ovarian tissue freezing.

She chose to proceed with ovarian tissue cryopreservation while understanding that it is an experimental procedure with the potential for future use if the scientific possibilities advanced. Several births have now occurred worldwide through the transplantation of this frozen ovarian tissue. Unfortunately, this tissue after transplantation tends to function for a rather short period of time, anywhere from nine months to three years, so its effects are not long-term.

❁ PRESERVING MALE FERTILITY

In the case of a male facing cancer treatments, his fertility can also be affected. Cancer treatments, including chemotherapy, radiation, and surgery, can affect a man's sperm and cause infertility. In fact, these treatments may affect not only the amount of sperm that a man can produce, but also the quality of the sperm. In addition to having a low sperm count, having slow moving or abnormal sperm can affect a man's fertility.

The best way to preserve a man's fertility is to collect a semen sample prior to radiation or chemotherapy. In some cases, a surgical procedure is needed to collect sperm directly from the testes. Sperm can be frozen and used in the future for intrauterine insemination or in vitro fertilization. Even if a man's sperm count is low or there are other concerns, banking sperm before starting treatment is a good back-up plan in case his fertility is affected by his cancer treatment.

Not all cancer treatments affect fertility the same way for everyone. It's important to talk with your doctor about how your course of treatment might affect your fertility. If you've already started treatment, keep in mind that some of these fertility options may still be available even after cancer treatment. Additionally, there are many other ways to start a family, such as using donor eggs, embryos, or sperm. Surrogacy or adoption may also be feasible options.

I recommend meeting with a reproductive endocrinologist who specializes in the unique fertility issues faced by cancer patients to help you navigate your many options for parenthood after cancer therapy.

Preserving fertility ranks as one of the greatest concerns women and young girls have when faced with a cancer diagnosis. Your cancer physicians are aware of this concern and they are willing to help you decide what would be best for you and the protection of your fertility. At CCRM we offer patients diagnosed with cancer expedited fertility preservation services.

CCRM's streamlined triage system allows patients to be seen by a physician within twenty-four to forty-eight hours of referral. CCRM also offers a random start stimulation approach to expedite clinical care. The ovarian stimulation and egg retrieval process takes less than two weeks.

We work with several non-profit organizations and pharmacies to provide enhanced and affordable access to fertility medications for our patients already facing overwhelming stress due to a cancer diagnosis.

It is scary to receive a cancer diagnosis, especially if you think it may threaten your ability to have children one day, but with the rapid advance in oncofertility and other areas of science, your chances of surviving cancer and having children are better today than ever before.

CHAPTER NINE

SINGLE EMBRYO TRANSFER AS THE SAFEST CHOICE

BY ERIC S. SURREY, M.D.

Patricia came to our clinic as a married thirty-seven-year-old in good health. She had never been pregnant despite trying to conceive with her husband, Michael, for three years. We evaluated her with ovarian reserve testing, ultrasound, and examinations of her uterus and fallopian tubes using both X-ray (hysterosalpingogram) and a scope (hysteroscopy). Everything was normal.

We also evaluated Michael and found him to be in good health generally. He did have an extremely low sperm concentration, however. His urologic and hormonal evaluations were normal. Based on our findings and discussions with the couple, they decided to try conceiving through in vitro fertilization (IVF) with the injection of a single sperm into the egg, also known as ICSI.

Patricia and Michael expressed concerns, however, about the traditional IVF process in which multiple embryos are transferred to maximize the chance for conception. While eager to conceive, they were hesitant to use multiple embryos because of a friend's difficult experience.

Patricia's best friend had undergone IVF with multiple embryos transferred and then had to stay in bed for two months because she became pregnant with twins and went into preterm labor. She then delivered the twins prematurely at thirty-three weeks. Patricia and Michael wanted to avoid this situation if possible.

I assured them that it was indeed possible to avoid these complications by successfully transferring just one embryo instead of several. This has become the recommended procedure in the last few years for younger, healthy patients.

It is only natural that patients planning to go through IVF to achieve pregnancy want to do everything they can to increase the odds of conceiving. In the past, this has meant transferring multiple embryos, particularly for women older than thirty-five.

Yet, as Patricia learned from her friend's experience, transferring multiple embryos greatly increases pregnancies with twins, triplets, or more. The more babies inside the womb, the more complications are likely to arise. These can threaten the health of both the mother and the children. That is the reason Patricia and Michael are not alone in their desire to achieve a pregnancy and successful delivery without using multiple embryos.

❋ THE RISKS OF MULTIPLE BIRTHS

While many infertility patients who've struggled to conceive are quite happy to go home from the hospital with an instant family or "two for the price of one," there is an increasing awareness that carrying more than one child increases the chance of complications including greater risk of premature delivery, which brings its own challenges as Patricia's friend experienced.

Studies have shown that gestational age at delivery is reduced by three weeks for each additional fetus. Also, researchers have found that twins or other multiple births pose significant other risks for both the mother and children. Complications can include pre-eclampsia, postpartum hemorrhage, maternal mortality, miscarriage, gestational diabetes, and pre-term labor.

The incidence of neonatal intensive care unit admission is increased by 48 percent for twins and by 78 percent for triplets or more. There is a fivefold increase in intrauterine death and a sevenfold increase in neonatal mortality for twins compared to single fetuses. Even after a successful delivery, infant mortality is nearly five to ten times higher for twins and triplets than for single infants. The incidences of cerebral palsy and severe handicaps are also significantly greater.

The risks for mothers are also increased by multiple births. The complications can include miscarriage, preterm labor and delivery, hemorrhage, pre-eclampsia, gestational diabetes, and operative delivery. In fact, it has recently been shown that maternal and neonatal outcomes were much better for women who had two IVF singleton pregnancies than one IVF twin pregnancy.

❋ ONE IS BETTER THAN TWO

When we first became capable of creating "test tube babies" through early IVF methods in the late 1970s, the most common approach was to transfer multiple embryos. The philosophy was "more is better" because the overall goal was to increase the odds for achieving pregnancy and a healthy baby, but success rates were extremely poor.

Then, as IVF success rates improved over the years, we focused more and more on finding ways to decrease the risks of multiple birth pregnancy without reducing the chances of conceiving. Finding that balance took years of research and testing. We have accomplished that goal thanks to medical advances like blastocyst culture and comprehensive chromosome screening (CCS), which has made us more adept at identifying the best embryos to transfer to the uterus—those that will most likely result in pregnancy.

This has been made possible by a big advancement, the ability to culture a human embryo beyond the old limit of two or three days to a more developed five-day (blastocyst) stage. These more mature embryos have many more cells. Because of this, we can evaluate them more easily to select the healthiest and most viable.

This is an important advance because only two of five eggs in young women have normal chromosomes, and that number decreases as women age. We've found that the more developed blastocyst embryos have a higher chance of implanting in the uterus, so fewer are required to achieve pregnancy.

CCS has been a major breakthrough for women over thirty-five. In the past, it has been necessary to transfer more embryos to achieve pregnancy in older women because the incidence of chromosomally abnormal embryos (which typically will not implant—or may result in early losses) dramatically increases with older women and potentially older men as well. However, once a CCS-tested embryo predicted to have a normal number of chromosomes is transferred, the likelihood of implantation is, on average, approximately 65 percent regardless of maternal age, according to our findings at CCRM.

However, once a CCS-tested embryo predicted to have a normal number of chromosomes (euploid) is transferred, the likelihood of implantation for every euploid embryo is, on average, approximately 65 percent regardless of maternal age, according to our findings at CCRM.

Others have confirmed these findings. We concluded then, that combining CCS with blastocyst stage embryo transfer significantly increases the success of eSET regardless of the woman's age. Since there are still women who do not conceive after transfer of a single CCS normal blastocyst stage embryo, there is still more research to be done to explain these failures.

Finally, single embryo transfer could not be accomplished successfully without advances in embryo freezing which has centered around the development of a successful rapid embryo freezing, or vitrification, program. Since instituting this technique at CCRM Colorado, our survival rates of embryos after vitrification has approached 99 percent with pregnancy rates that are not significantly different than those obtained with fresh embryo transfers.

Other exciting areas of research include evaluating the metabolism and gene expression of individual embryos by creating a profile of the culture media in which they have been allowed to develop. This research, some of which has been performed in our laboratories, is not yet ready for clinical use, but offers new ways to select the single most viable embryo for transfer. The potential for the creation of "fingerprints" of individual embryos that include an assessment of development, chromosomes, metabolism, and gene expression looks promising and may allow for routine single embryo transfer in all patients.

❁ THE RESEARCH

Logically, you would assume that the more embryos implanted during IVF, the greater the likelihood of conceiving, but studies have shown that isn't really the case. One early study found that when more than two embryos were transferred to healthy women under thirty-five with a good chance of pregnancy, it did not improve pregnancy rates, but it does result in a significant number of more triplets.

As a result of this and similar findings, in 1999 the Society for Assisted Reproductive Technology (SART) proposed a reduction in the number of embryos to transfer. It said that no more than two should be transferred in this younger group of women. However, this guideline still allowed for the transfer of up to four embryos for women thirty-five to forty years old. Since then, additional research has resulted in the SART guidelines becoming more and more conservative in regard to the number of embryos transferred. In 2013, the organization recommended that no more than one or two embryos should be transferred for women thirty-five to forty years of age.

Fertility specialists have followed these guidelines and we've seen significant decreases in the incidence of high order multiple pregnancies even as live birth rates have increased. Given that reduction of the rate of twins was not the goal of this initial change in guidelines, the incidence of twin pregnancies has not really changed.

While it was becoming clear that transferring just one embryo was an option worth considering because it reduced the risks associated with multiple pregnancies, there was still the question of how to do that without also reducing patients' chances for a pregnancy. Researchers found that using single embryos allowed to mature only three days cut down on multiple pregnancies, but also lowered pregnancy rates.

That wasn't an acceptable option to patients eager to have a child. One approach might have been to do more than one cycle of frozen embryo transfers, but that also could bring additional cost for the patient as well as added emotional trauma for each unsuccessful transfer.

A more reasonable, and quite effective solution was to allow embryos to mature longer in the laboratory, for five instead of three days to the blastocyst stage. These more mature embryos have a better chance of implanting in the wall of the uterus.

While the day three (cleavage stage) embryo has only six to ten cells, the day five (blastocyst stage) embryo has sixty to one hundred cells so it is more highly developed. It is thought that in natural conception, the embryo does not actually enter the uterus from the fallopian tubes until it reaches the blastocyst stage at five days after conception, so that transfer of the blastocyst embryo in IVF is actually more of a natural process.

In one randomized trial by European researchers, they found that transfer of a single blastocyst produced pregnancy rates of 37.3 percent compared to only 24 percent for transfer of a single day 3 embryo.

We took it a step further in our own randomized study at CCRM. We transferred a single blastocyst embryo in one group of healthy women in the thirty-five-year-old age range and in the other group of same aged women we transferred two blastocyst stage embryos.

The single blastocyst group had a pregnancy rate of 61 percent while the two blastocyst group had a 73 percent pregnancy rate, which isn't a major difference statistically. There was a major difference, however, in the rate of multiple pregnancies between the groups, however. It was found that

47.2 percent of the women who received two blastocysts became pregnant with twins, as opposed to none of the women who received a single blastocyst.

As a result of these and other studies, the American Society for Reproductive Medicine (ASRM) further revised its guidelines for numbers of embryos to transfer in 2013. The group advised replacement of no more than one blastocyst stage embryo to women under thirty-five who had not previously failed IVF cycles and had good embryo quality—and to those women receiving embryos derived with the assistance of an egg donor.

The new guidelines still recommend transfer of one to two blastocysts for women thirty-five to forty and three blastocysts for those over forty—even those with a good prognosis.

❀ *THE SINGLE EMBRYO OPTION*

Although more and more patients have chosen single blastocyst embryo transfers since it became an accepted method, SART reports that in 2015 this option was selected by only 41.6 percent of women under thirty-five years of age and just 31.8 percent of those women in the thirty-five to thirty-seven year age group.

Patients still have been a bit reluctant to choose single embryo transfer despite its advantages because of the desire to maximize opportunities for successful pregnancies and delivery, as well as economic concerns about the cost of undergoing multiple transfers and the emotional toll of failed cycles. In a Dutch study, close to half of women were open to single embryo transfer if pregnancy rates were unaffected, but if the rate dropped by as little as 3 percent, only 24 percent were open to the idea. In another study from Denmark,

58 percent preferred having twins compared to only 37.9 percent who preferred having one child at a time.

Some western European countries have mandated elective single embryo transfer by law. Although this approach would obviously expand the use of eSET and decrease multiple pregnancy rates, overall pregnancy rates per transfer have been compromised as a result. Given that IVF in these countries is paid for by national health systems, acceptance is not a matter of choice, but of legislation.

Certain clinics in the U.S. have instituted a mandatory single blastocyst transfer policy for good prognosis patients who are younger than thirty-eight. Multiple birth rates reportedly have dropped dramatically while live birth rates remain high. This approach does not address the issue of patient autonomy.

❁ *RECOMMENDED CANDIDATES FOR ELECTIVE SINGLE EMBRYO TRANSFER (ESET)*

It is important to note that not all problems with pregnancies after IVF are solved with eSET. There is still a risk of identical twins, which appears to be increased after IVF. This has been reported to range from 1.2 percent to 8 percent in comparison to 0.4 percent in all births. A recent literature review also showed that eSET is associated with decreases in preterm birth and low birth weight in comparison to two embryo transfers. However, the risk of preterm birth remained higher than after natural conception.

In selecting the best candidates for elective single embryo transfer today, we recommend the following:

- women under 35 percent without prior unsuccessful IVF cycles who are undergoing blastocyst stage embryo transfer

- participants in egg (oocyte) donation cycles with high quality embryos

- those with a history of obstetrical complications such as preterm labor or birth, pre-eclampsia, severe hyperemesis (intractable nausea and vomiting), and/or gestational diabetes

- women with medical conditions that could be made significantly worse by pregnancy by impacting fetal and maternal well-being such as hypertension, diabetes, or kidney, heart, or lung disease

- those with congenital uterine abnormalities

Finally, I think the case could be made to use eSET for women of any age with good quality blastocyst stage embryos expected to have a normal number of chromosomes after CCS. In a very short period of time, IVF has progressed from a controversial therapy with limited success to a highly effective and well-accepted approach to the management of infertility. The one caveat is the high incidence of multiple pregnancies.

Advances in IVF laboratory techniques such as blastocyst culture, CCS and vitrification have made it possible to drastically reduce the risks of multiple pregnancies without sacrificing the chance of a successful cycle.

Exciting research may allow us to evaluate simple secretory products and gene activity simply by analyzing the culture medium in which the embryo has been developing. This could provide even more information to help with selection of the best embryos for transfer.

These advances may open the door to routine elective single embryo transfer for all. The ultimate goal for all IVF patients is that one day they each will routinely give birth to a single child for each single embryo transfer procedure.

Distribution of Multiple-Fetus Pregnancies and Multiple-Infant Live Births Among ART Cycles Using Fresh Nondonor Eggs or Embryos, 2015

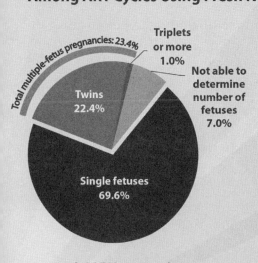

Total multiple-fetus pregnancies: 23.4%

Triplets or more 1.0%

Not able to determine number of fetuses 7.0%

Twins 22.4%

Single fetuses 69.6%

A. 26,708 pregnancies

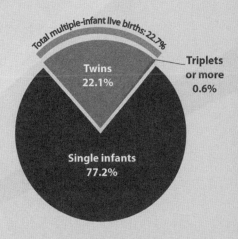

Total multiple-infant live births: 22.7%

Triplets or more 0.6%

Twins 22.1%

Single infants 77.2%

B. 21,771 live births*

* Total does not equal 100% due to rounding.

National Center for Chronic Disease Prevention and Health Promotion
Division of Reproductive Health

(source: CDC.gov)

● *CANDIDATES FOR ESET:*

- Under thirty-five years old without prior IVF failure and good blastocyst quality

- donor oocyte recipient with good blastocyst quality

- any age patient after CCS with good blastocyst quality

- congenital uterine abnormalities

- prior obstetrical complications (including preterm labor, preterm birth, pre-eclampsia, severe hyperemesis, gestational diabetes)

- high risk medical conditions (including hypertension, diabetes, kidney, cardiac, pulmonary diseases)

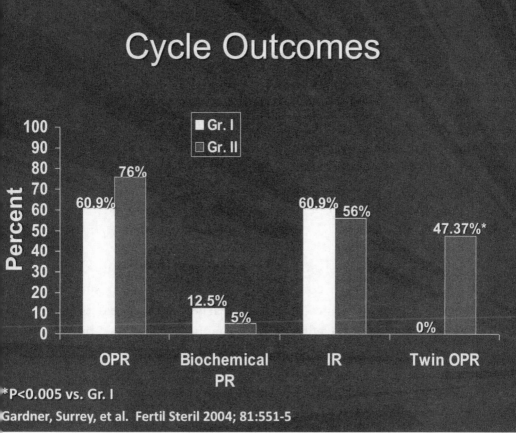

Figure Legend: Likelihood of ongoing pregnancy (left) and twins (right) after transfer of one (shaded) versus two (white) blastocyst stage embryos at CCRM. From Gardner et al. Fertil Steril 2004; 81: 551-5.

CHAPTER TEN

UTERINE EVALUATIONS REDUCE RISK

BY DOCTOR ERIC S. SURREY, M.D.

Emily, a thirty-six-year-old, and her husband, Mark, had been trying to conceive for five years without success, including a pregnancy that failed early on just a year before they came to our clinic. Their diagnosis was one that is especially frustrating: "unexplained infertility." No one had been able to nail down any particular cause for their inability to have children.

Ovarian reserve and sperm function testing had been normal. The X-ray examination of the uterus, fallopian tubes, and surrounding areas (hysterosalpinogogram) showed both tubes were open and the uterine cavity "appeared normal." The baseline ultrasound examination was unremarkable with the exception of a report of a "slight irregularity" in the back wall of the uterus.

Emily had undergone three cycles of treatment with Clomid, a drug commonly used to induce ovulation, and intrauterine insemination (IUI) resulting in the one pregnancy loss. An IVF cycle performed at another clinic resulted in the transfer of two good quality blastocysts, but they did not lead to success.

As part of her evaluation at our clinic prior to attempting another IVF cycle, we performed an office hysteroscopy, in which a thin flexible tube known as a hysteroscope is inserted through the vagina into the cervix and then inside the uterus. The procedure revealed a 1.5-centimeter fibroid that was distorting the back wall of Emily's uterine cavity. Although we cannot say that the fibroid was the sole cause of Emily's infertility, it certainly was a contributing factor as we shall discuss further in this chapter. The fibroid was not large. The bigger issue was its location in the uterine cavity. The question we had to ask was: "How was this missed during the previous evaluations?"

We generally recommend that all patients who plan on becoming pregnant one day undergo a thorough evaluation of the uterine cavity, especially those who are undergoing an embryo transfer. We see this as a wise preventive measure that should be done rather than risk going through an IVF cycle that fails, which can be traumatic emotionally and expensive.

Emily's case offers a good example of why we make this recommendation. After discussion with her physician, Emily underwent outpatient hysteroscopic resection of the fibroid. A follow-up office hysteroscopy performed two months later revealed a normal uterine cavity with no evidence of scar tissue. She underwent an IVF cycle with transfer of a single, excellent quality blastocyst stage embryo three months later. She conceived and delivered a healthy daughter at term without complications.

❀ METHODS FOR EVALUATING THE UTERINE CAVITY

Failed implantation of the embryo in the uterus is a common cause of unsuccessful IVF cycles, and possibly, infertility in general. We believe that a thorough evaluation of the uterine cavity should be an integral part of preparation for all IVF patients as well as gestational carriers (surrogates). The current gold standard for assessing the uterine cavity is hysteroscopy. With this method, a thin scope is inserted into the vagina and moved gently into the cervix and then the uterus. We then distend the uterine cavity with either carbon dioxide or a sterile clear fluid solution. The hysteroscope has a light and a camera at the end so we can clearly see those areas on a monitor. If we think it will be helpful, we take a small sample of tissue to perform a biopsy.

In the past, this procedure could only be performed in an operating room with the patient under anesthesia and required cervical dilation. However, now that thin, flexible fiberoptic hysteroscopes have become available, the overwhelming majority of these diagnostic procedures can be performed in an office setting requiring only local or typically no anesthesia and without cervical dilation, to minimize the discomfort.

Emily's case provides an excellent example of why hysteroscopy is widely considered a more sensitive method for evaluating the uterine cavity than either the X-ray examination, hysterosalpinogogram or HSG, or the baseline ultrasound exam. As I noted, in Emily's examination the HSG showed both tubes were open and the uterine canal appeared normal. The baseline ultrasound examination was unremarkable except for a slight irregularity in the back wall of the uterus. The hysteroscopy showed that the "slight irregularity" was a fibroid.

As you can see in this patient's situation, the correlation between hysteroscopy and hysterosalpingograms (HSG) or baseline ultrasound examinations is extremely poor. In one study, the sensitivity of HSG was only 21.6 percent while agreement between HSG and hysteroscopy findings was only 68.9 percent. Baseline ultrasound examinations do not fare much better.

Diagnostic hysteroscopy was recently shown to be significantly more sensitive (97.26 percent vs. 89.04 percent) and specific (97 percent versus 56 percent) in detecting uterine abnormalities than baseline transvaginal ultrasound exams. That is why uterine abnormalities had not been completely ruled out before Emily came to us, even though she'd reportedly had normal HSG and only inconclusive findings on baseline ultrasound exam in her previous examinations at other clinics.

In one study, researchers looked at a group of patients of similar ages who had normal HSG examinations, but also underwent hysteroscopy prior to an IVF cycle. Dramatically lower pregnancy rates were reported in those with abnormal findings at hysteroscopy who also had normal findings at HSG when compared to those for whom both studies were normal (8.3 percent versus 37.5 percent). Another group showed that 22 percent of women planning IVF had abnormalities on hysteroscopy, but normal HSG and/or baseline ultrasound exams.

In another telling report, women who had failed at least two prior IVF cycles with transfer of good quality embryos and a normal uterine cavity as described by HSG, then underwent hysteroscopy. Forty-five percent had abnormal findings. Once these abnormalities were corrected, the clinical pregnancy rate in the next IVF cycle was 52 percent.

❂ ALTERNATIVES TO HYSTEROSCOPY

There are alternatives to hysteroscopy that may achieve a similar degree of accuracy.

Saline infusion sonohysterography (SIS) is an office-based technique in which a sterile saline solution is infused into the uterine cavity through a small catheter at the same time that transvaginal sonography is performed. By distending the uterine cavity with the saline solution, visualization is significantly enhanced. Studies have shown that SIS and hysteroscopy are generally equivalent in their ability to diagnose uterine cavity abnormalities.

Three-dimensional (3D) enhanced ultrasound technology can further improve the accuracy of SIS. In one study, both 3D enhanced SIS and outpatient hysteroscopy were equally helpful and both were superior to more traditional 2D SIS in diagnosing uterine abnormalities.

In general, MRI (magnetic resonance imaging) is an extremely costly procedure that should not be considered as a primary diagnostic tool. There is one exception, however. This approach may be extremely helpful in the diagnosis of adenomyosis, or congenital uterine abnormalities, which I will cover later in this chapter.

❂ TYPICAL UTERINE ABNORMALITIES AND THEIR TREATMENTS

The most common uterine abnormalities that may impact fertility, implantation or pregnancy loss include:

Fibroids:

- Uterine Fibroids—benign smooth-muscle tumors that are extremely common in reproductive age women.

Fibroids are described by their location, but the makeup of these tumors is the same.

- Subserosal fibroids—fibroids that distort the outside of the uterus and may cause pressure symptoms, constipation, or urinary frequency depending on their size and location. These are rarely associated with reproductive difficulties unless they are extremely large. Surgical removal through the abdomen by laparoscopy or an open incision is primarily warranted for symptom relief only.

- Submucosal fibroids—typically smaller and grow into the uterine cavity. These tumors can cause abnormal uterine bleeding, infertility, and recurrent pregnancy loss, but may not cause any symptoms. Removal of these types of fibroids may be accomplished in a minimally invasive fashion under anesthesia using operating hysteroscopes that have special loop attachments that allow for safe removal of the fibroids through the cervix. We have published the results of a study in which we found that performing these procedures prior to embryo transfer resulted in ongoing pregnancy rates after IVF that were similar to those of matched controls without fibroids at all.

- Intramural fibroids—located within the uterine wall, their impact on infertility and the pros and cons of surgical removal on fertility or pregnancy loss is more controversial. We have reported that even when hysteroscopic findings were normal, the chance of an embryo implanting is lower in the presence of intramural fibroids in women less than forty years of age. When those intramural fibroids that are adjacent to or distort the uterine cavity are removed surgically, the resulting pregnancy rates are similar to those achieved in women without fibroids as well.

Uterine Polyps

Polyps represent one of the most common abnormalities found in the uterine cavity. Polyps are typically benign growths composed primarily of glandular cells. They have been associated with abnormal bleeding, infertility, and rarely, malignant change in older women. There is controversy regarding the effect of these lesions on IVF outcome. There are very few studies that evaluate the impact of hysteroscopic removal of polyps prior to IVF. Some have shown increased pregnancy and decreased miscarriage rates, while others have shown no impact. It has been a general consensus that polyps greater than one centimeter in mean diameter, and those causing symptoms should be removed. The impact of smaller polyps is less clear, but removal should be considered if they are symptomatic, appear to be growing, are multiple, or are found in patients with unexplained prior IVF cycle failure.

Congenital uterine abnormalities

A variety of congenital uterine abnormalities has been identified. Most are associated with recurrent pregnancy loss as well as preterm labor and birth, but there is little evidence to suggest that they cause infertility.

- Uterine Septum—a wall in the middle of the uterine cavity

- Bicornuate—a malformed uterus in the shape of a heart

- Didelphys Uterus—a rare condition also known as a "double uterus," this is a malformation that occurs when the two tubes that normally join to create the uterus do not come together so each creates a separate structure. A double uterus may have just one opening (cervix) into one vagina, or each uterine cavity may have its own cervix. There may even be two vaginas.

- Unicornuate Uterus—a malformation in which the uterus is formed from one of the normally two paired Müllerian ducts, while the other Müllerian duct does not develop or develops only in a rudimentary fashion

A hysteroscopy or HSG alone is not sufficient to conclusively diagnose these abnormalities. An MRI or 3D ultrasound helps us differentiate one abnormality from another. This is particularly helpful with regard to septate vs. bicornuate uterus, which can look similar with both hysteroscopy and HSG, but can be fairly easily differentiated by 3D ultrasound or MRI. A uterine septum can be repaired fairly easily with hysteroscopic surgery, while the other malformations above cannot typically be repaired.

Any patient with a congenital uterine anomaly should have an imaging study performed of her kidneys and ureters because there is a high correlation with congenital urinary tract anomalies including absent kidneys, pelvic kidneys, and abnormally positioned ureters, particularly in those women with a bicornuate, didelphys, or unicornuate uterus.

One of the ongoing controversies among reproductive surgeons is whether surgical repairs should be attempted on a patient with a uterine septum who is planning IVF but has no history of recurrent pregnancy loss or preterm birth or delivery. One study has shown that compromised pregnancy rates after embryo transfer were no different for a controlled group than for women with a uterine septum after surgical repair. However, there are no randomized studies that address this subject.

Uterine Adhesions

Uterine adhesions or scar tissue typically results from prior uterine infection or surgery, or retained placental tissue from a previous pregnancy. Scar tissue that forms after a post-partum D&C (dilation and curettage to remove tissue from inside the uterus) is called Asherman's syndrome. Hysteroscopy and, to a lesser extent, 3D saline infusion sonohysterography (SIS), are the best means for making this diagnosis. Adhesions can typically be removed through hysteroscopic surgery, but more extensive scarring may require more than one procedure, placement of a post-operative intrauterine balloon catheter for several days and/or administration of a course of estrogens to promote healing. In the worst cases in which the cavity cannot be normalized, the only alternative is to consider gestational surrogacy.

Adenomyosis

Adenomyosis is a disorder in which endometrial glandular tissues migrate to the muscle layer of the uterus. This is a relatively common finding noted in 20–30 percent of the general population and is often symptom free. However, it can also be associated with severe pelvic pain, abnormal uterine bleeding, and painful periods and intercourse. A recent analysis of previously published papers evaluating the impact of adenomyosis on IVF outcome reported a significant reduction in clinical pregnancy rates and increased miscarriage rates in women with adenomyosis in comparison to those without.

This disorder is best diagnosed with 3D-ultrasound and/or MRI. Hysteroscopy is not particularly helpful in this circumstance. Treatment, however, is somewhat problematic for women who wish to conceive since hysterectomy has typically been the definitive surgical therapy. Clinically significant isolated pockets of adenomyosis, called adenomyomas, can often be removed surgically.

If adenomyosis is more extensive, conservative surgery is rarely successful, although limited success has been obtained by administering a prolonged (two- to three-month) course of a medication (GnRH agonist), which suppresses the production of estrogen creating a temporary menopausal state to stop menstruation and potentially shrink the adenomyosis prior to an embryo transfer cycle. Some have suggested placing a progesterone-containing IUD for three to six months prior to embryo transfer as an alternative. There is minimal supportive data for either of these approaches. If these manipulations are unsuccessful, gestational surrogacy may represent the only viable alternative.

❀ *WAITING TIME AFTER UTERINE SURGERY*

The ideal time interval after uterine surgery to attempt pregnancy has not been well established and varies with the type of surgery. The longest delay should occur after open abdominal myomectomy, in which fibroids are removed through an incision in the lower abdomen, after laparoscopic myomectomy. Most surgeons would recommend a three- to six-month wait before attempting an embryo transfer depending on the complexity and extent of the procedure. Caesarian delivery may be necessary if the incisions in the uterus enter or come close to the cavity.

With regards to hysteroscopic surgery, a recent study from Taiwan noted that the uterus was typically well-healed one month after removal of a uterine polyp or intrauterine adhesions. However, two months were necessary to achieve healing after a septum was resected and two to three months were required after hysteroscopic resection of a submucosal fibroid. This underscores the importance of a "second look" office hysteroscopy after these procedures to document appropriate healing and to rule out the presence of new or residual scar tissue.

❂ OTHER UTERINE TESTS TO IMPROVE IVF OUTCOME

There are a variety of substances that play a role in the implantation process, some of which involve "cross-talk" between the uterus and the embryo. A very small number of these substances can be evaluated after appropriately timed biopsy of the uterine lining or potentially from the fluid secreted into the uterine cavity. However, given our lack of complete understanding of the implantation process, we are presently unable to paint a complete picture of the causes of implantation failure. Thus, the tests described below are considered experimental at this time, are in no way complete and should not be used on a routine basis. I just wanted to share them with you so you can see what may be on the horizon.

Several investigators have shown that the absence of expression of the protein ß3 integrin, which is involved in implantation, is more common in women with endometriosis, hydrosalpinx (tubes blocked at the far end), and those with unexplained failed embryo transfers.

Administering the same GnRH agonist (described above) to women missing this protein for two to three months prior to an embryo transfer has resulted in improved pregnancy rates by helping to restore expression of the protein, according to some reports. Others have disputed these findings, however. There have been preliminary studies suggesting that evaluating endometrial leukemia inhibitor factor along with integrin expression may be more valuable than evaluating the presence of integrin alone, although the data is sparse.

A commercially available "endometrial receptivity assay," which evaluates the expression of certain proteins in the uterus that effect a "window of implantation," has been reported to be helpful in better predicting the appropriate number of days of progesterone exposure prior to frozen embryo transfer or fresh transfer in an egg (oocyte) donation cycle. Although this is extremely intriguing, there is currently a paucity of outcomes data (i.e., pregnancy) to support its use at the time of this writing.

Clearly, this avenue of investigation is relatively new and with advances in our understanding of the implantation process, new means of evaluating uterine gene expression and secretory products hold great promise for not only diagnosing, but also treating unexplained implantation failure.

The health of the uterus represents a critical component in achieving a successful pregnancy. Although there are a small number of well-designed studies, evaluating the uterus in an accurate way clearly represents a critical part of the infertility evaluation. This allows the patient and her physician the opportunity to discuss the pros and cons of treatment options with a maximal amount of information. There is no reason to allow a patient to undergo a failed cycle that could have otherwise have been prevented before considering the testing discussed in this chapter.

Techniques for evaluating the uterine cavity:

- Hysteroscopy

- 3D-enhanced saline infusion sonohysterography (SIS)

- MRI (particularly for adenomyosis and congenital uterine abnormalities)

- Less sensitive:

- 2D-saline infusion sonohysterography (SIS)

- Least sensitive:

- Baseline ultrasound examination

- Hysterosalpingogram (HSG)

- Pelvic examination

- Most common uterine abnormalities:

- Fibroids (leiomyomas)

- Polyps

- Congenital abnormalities:

- Septum

- Bicornuate

- Unicornuate

- Didelphys

- Müllerian agenesis (absent uterus, fallopian tubes, cervix, top third of the vagina)

- Adhesions (scar tissue)

- Adenomyosis

CHAPTER ELEVEN

MIND & BODY FERTILITY TREATMENTS

BY JUDITH J. BECERRA AND KIM KLUGER-BELL OF THE CCRM COUNSELING TEAM

Emily was in her mid-thirties when she and her husband, Jason, decided to start their family. Initially, Emily couldn't get pregnant, so she sought treatment with a local clinic. They found she had polycystic ovaries and her doctor put Emily on a fertility drug. Within a year, she still had not become pregnant.

The couple had more difficulty when they tried a more aggressive approach. Emily had a miscarriage after twelve weeks, multiple failed medicated cycles, three failed attempts at intrauterine insemination (IUI), and a chemical pregnancy, before she came to our clinic for assistance.

She and her husband were primarily focused on in vitro fertilization to help them have a child. Emily discovered that our staff also includes counselors who help patients deal with the emotional and spiritual challenges that infertility challenges can cause.

As a clinical social worker, Emily assumed she was well prepared to manage the feelings of frustration, helplessness, and anger that often arise for our patients. However, she came to appreciate the value of our support groups and the coping methods we suggested. "When we were having so much trouble trying to have a child while also working hard at our careers and trying to have a normal life," she said, "I had anxiety and stress and depression building on top of one another."

Fortunately, we were able to help Emily accomplish her goal through IVF. Before she came to us, she had struggled for two and a half years to maintain a full-term pregnancy and delivery. Their son, Dalton, was born in late 2014.

✿ *AN UNBEARABLE BURDEN*

Throughout her journey with us, we encouraged Emily to try a number of approaches for alleviating stress and managing negative feelings. Her experiences have their unique aspects, certainly, but nearly all women who have difficulty getting pregnant experience similar emotional challenges.

Infertility has emotionally crippled the lives of women throughout history.

As early as in the Old Testament of the Bible, there are stories of brokenhearted women who are "barren" and weeping in anguish. It also depicts men pleading with the Lord for their wives to conceive. Anxiety, heartbreak, and even shame have added to the burden of women and men dealing with infertility over the generations. Unfortunately, there hasn't been nearly as much emotional support as medical support for patients. Often, they've felt like they could not share their feelings with those closest to them because of the intimate nature of this medical problem.

It's also true that until recent times, there have not been many places to turn to for professional guidance. The counseling staff members at CCRM have made it their mission to change that. We are strong advocates of a counseling approach that addresses the needs of both mind and body during infertility treatment. This approach emphasizes the mind-body connection by acknowledging that what you think, how you feel emotionally, and what you believe to be true directly affects your health, and may also affect your fertility.

Our goal as counselors at CCRM is to reach out and let patients know about the support services we offer. We have support groups meeting weekly at each of our sites and many of our patients find that they become a critical part of their treatment, a source of comfort, and a refuge from their emotional pain.

Both of the co-authors of this chapter went through infertility over twenty years ago when the emotional aspects of infertility were not often discussed or treated. Although our fertility doctors were extremely kind and brilliant physicians, they were not trained to deal with patients' feelings. Fortunately, most doctors today understand the importance of the mind-body connection and the many ways emotional stress can impact a patient's physical health. We feel privileged to work at a clinic whose staff not only validates, but also embraces, this mind-body approach to fertility.

Thanks to the groundbreaking work of the Mental Health Professionals Group of the American Society of Reproductive Medicine as well as psychologists such as Harvard's Alice Domar, a pioneer on the psychological aspects of infertility treatment who founded the first mind-body clinic for women, we now have a much better theoretical framework in which to serve our patients' mental and emotional needs.

Our clinic is among the few that provides patients with mental health counselors during their treatment that are imbedded in the clinic. We work with the medical team to help patients manage the mental and emotional challenges of infertility. The women who enter our center are not seen merely as patients who need medical treatment; we see them as fellow human beings with complex thoughts, feelings, relationships, and lives outside of the fertility center.

❀ *CHALLENGES TO THE MIND, BODY AND SPIRIT*

In an age where people freely discuss—on social media, reality television, and in public—nearly any subject, no matter how personal or private, there still seems to be shame and secrecy associated with infertility. This explains why many of our patients come to us feeling stressed and alone in their struggle. They often have been humiliated and hurt by insensitive comments, a lack of understanding, and a feeling that others are somehow blaming them for their infertility. Too often, couples facing infertility issues isolate themselves as a result.

Emily learned that the world can be a lonely place when you are infertile. She found herself avoiding social situations and conversations about children and pregnancy. Friends and acquaintances were divided among "the fertiles" and the "infertiles," the "moms" and the "non-moms."

"Eventually, we stopped going out when we realized that the fertiles of the world would never understand what we were going through," she shared.

Like Emily, many patients tell us that they feel they can't talk with family and friends about this major challenge. We do our best to be there for them, of course, and we also help them find support groups in their communities and online, including web forums and private Facebook groups. These are great sources of information and emotional support, especially since most are available any time of the day or night.

We share the belief that "A problem shared is a problem halved." Family members and friends may be empathetic, but there is nothing quite like talking to someone who has dealt with her own infertility issues, someone who knows what it is like to endure all of the tests and treatments as well as the physical and mental stress. Only someone who has been through this experience truly understands why you can feel grief when you see other happy mothers with their children, or what it feels like to have your body pumped full of hormones, or the agony of going through one failed pregnancy test after another.

While we encourage our patients to lean on family and friends rather than isolating themselves, we also understand that for a while at least, they might find it easier to share their stories, seek solace, and communicate with fellow infertility warriors online. Many of our patients say they have benefitted from their local and online support communities provided by RESOLVE, the National Infertility Association (www.resolve.org).

The internet can be a great resource, yet we advise our patients to be cautious about acting upon or accepting medical advice or information from online sources. There are many scam artists out there and they've become skilled at masking their intentions by posing as physicians or other medical experts. Some websites that extol the virtues of medications or natural remedies are disguised to look like news media or non-profit sites, but they are actually maintained by companies that stand to profit from selling those items. As we say in our office, "Google is not your doctor!"

Even information provided by other infertility patients might not be valid in your case because each patient is unique. What works for one patient might not work for another, and vice versa. Never accept any information from an online source until you have discussed it with your physician.

❀ *HELPING EACH OTHER*

Many of our patients find that their greatest sources of comfort and knowledge are local support groups that we help them find based on their specific fertility challenges. Our former patient, Emily, is among those who made lasting friendships with the other women and couples in her support groups.

She and her husband, Matt, whose story is told in greater depth in our egg donor chapter, struggled for two years to have children before coming to us. We found that although Emily was only twenty-eight years old, the condition of her eggs was like that of a forty-year-old.

After looking at all of the options available and weighing each of them, this couple decided to use an egg donor and then IVF to transfer two embryos. There were hurdles to overcome, but today Emily and Jason have beautiful twin daughters. They often return to our clinic to offer encouragement to other women and couples facing infertility and the emotional upheaval that can result.

"In the beginning, Emily wasn't talking to anyone about this and I wanted her to," Jason recalled. "I felt helpless. I couldn't hold her hand and fix it. I told her she needed to find a place where she could express how she was feeling."

Emily blamed herself for her infertility at first. After talking with counselors and other women who'd gone through a similar experience, she was able to free herself of that unwarranted burden.

"I was wallowing and had isolated myself, but that changed when I accepted that there wasn't anyone or anything specific at fault," she said. "Once I did that, and began discussing our infertility openly, we were able to move forward.

"I wish I had some perfect and magical answer, but I don't. All I can really say is that I am grateful to have had a supportive spouse as well as family and friends, including our donor egg support group. We wouldn't have pushed along as we did had it not been for them.

"Jason and I attended a general infertility support group at the Louisville CCRM location once, we also participated in RESOLVE support groups multiple times, and CCRM's donor egg support group on a monthly basis," Emily said. "The latter was the most helpful because it was specific to our infertility issue and I had finally found 'my people' if you will. They were a major part of my stress relief. The support we gave each other couldn't be matched or even exceeded by anything."

"Just knowing that other couples were facing the same issues and having the same emotional turmoil helped 'normalize' the challenge," Jason said.

"When we heard people in our infertility group talking about all of the miscarriages and problems with cancer and secondary infertility, it made our problem seem almost insignificant because even though we'd never had a positive pregnancy test, we hadn't lost a pregnancy, which was an experience so many others had gone through," he said.

Emily and Jason have remained close to four of the families they met in the donor egg program. Three of the others have twins too. Now, they are enjoying raising their children together, bonded by their shared experience.

"We were all pregnant at the same time and it was unreal to have other couples going through the same thing. I felt like I belonged with these other women and we are all still very close," she said.

Local support groups for our fertility patients often result in lifelong friendships. Sometimes, however, patients also ask for individual and couples counseling, which we also provide at CCRM. When requested, we can even go into the pre-surgery room or the transfer room in the surgery center and help patients who feel anxious. We have patients and couples who take advantage of this and we have contact with them every week when they are in treatment.

● STRONG OF MIND

Usually by the time patients come to us they are frustrated, angry, and afraid because of their failed efforts to start a family. Many of them are also fed up with well-intentioned advice like "just give it time," or "stay positive." While we encourage patients to stay positive, we know it's not easy to ignore negative thoughts when your fertility tests keep coming back with bad news. You should know that this is a natural tendency, and part of your instinctive emotional defense system. You are trying, subconsciously, to protect yourself from further disappointment, building a wall around your hopes and dreams.

Unfortunately, this isn't the most helpful approach, and there are more effective ways to overcome disappointments and frustrations. Instead of sinking into despair and giving up, you can tap into a more affirmative and proactive approach. The secret is to focus on controlling those things that are within your power to control.

There are many things you simply cannot control in your quest for fertility. Yet, you can take command of self-defeating negative feelings and shut them down. Be aware of your inner "Debbie Downer" and her power to drag you down with unproductive thoughts. Become the gatekeeper by stopping that nasty little voice whenever it crops up in your head. If you are familiar with the *Saturday Night Live* character of that name portrayed by actress Rachel Dratch, you might try visualizing her face saying "no" whenever negativity threatens. If that's a bit too theatrical, you might visualize a stop sign in your mind whenever negative thoughts creep into your mind.

Use whatever image works best for you. Some patients visualize the large red and white sign or a big hand indicating "Halt!" We also suggest that patients create an image of a storage box in their minds and use it to lock up self-defeating thoughts when they arise.

Some women find it helpful to first sift through the negative thoughts for a brief period, or even talk about them with someone. We recommend that they do this only for a brief time before stowing away the negative thoughts. This helps minimize the power the negative thoughts will have.

It's okay to be aware of the thoughts and to evaluate whether they have any merit, but you don't want to let them linger. Ask yourself, "Is this a helpful thought or feeling? Is this thought making me feel hopeful?" If the answer is no, shove it in the box and put it on the shelf in the back of your mental storage unit.

❁ *RUMINATION CAN LEAD TO RUINATION*

It is not uncommon for patients to fixate on the problems of infertility. Spouses and partners often complain that patients don't want to talk about anything else, that their desire to have children has become an obsession. This is natural and understandable, but rumination can be your ruination.

It's unhealthy to constantly focus on any one thing, whether it's work or infertility. Spouses and partners can become exasperated and withdrawn when every conversation is focused on this one, draining issue. It's also true that women undergoing treatment often feel that their partners and spouses shut down on them, and that passion diminishes.

"My husband and I did deal with my obsessiveness," Emily recalls. "I would do all of this research and spew it out to him. He would try to be supportive and assure me that we could afford treatments and that it would all work out. But I'd say, 'how can you be so sure?' We had a twelfth-week miscarriage and other problems!'"

Emily admits she was fortunate to have a husband who has a calm temperament. "He is very even-keeled and he never got upset with me, but I'm sure a lot of spouses would," she said. "He was content to not have children and I never worried that he would leave me if I couldn't get pregnant."

Like many women, Emily felt that some of the intimacy of their relationship suffered because their sex lives became more regimented and scheduled during treatment.

"It felt like we weren't as close because it was all about the clockwork and when I was ovulating," she said. "I was so focused on trying to make this happen. He would tell me to compartmentalize my concerns on getting pregnant, but it was in my thoughts twenty-four hours a day. I would obsess over it and I couldn't shut it off. I'd go to the park and see other women pushing baby strollers and it was crushing. I couldn't shut it off, even when I was pregnant, I'd worry about my blood work results and losing the baby."

Because fertility treatments regiment a couple's sex life and require physical intimacy "on demand," many couples experience a diminishment in their sexual desire for one another. Add to that a single-minded focus on the achievement of the goal of a positive pregnancy test, and it can create an even further distance between a patient and her partner.

❁ WAITING FOR THE NUMBERS

Obsessive worry is all too common for our patients. We try to help them alleviate this tendency as much as possible. Emily's husband was giving her good advice when he told her to try to compartmentalize her worries and concerns about getting pregnant. We tell patients to set aside a limited amount of time each day to talk with spouses and loved ones about infertility and the emotions they are having. Once the time limit is up, put those matters aside and talk about more upbeat and everyday things. This helps keep a balance and, more importantly, reduces tension and stress that can be a barrier to fertility.

Many patients undergoing fertility evaluations and treatment say that the most nerve-wracking periods—and the times when negative thoughts tend to overwhelm them—are those spent waiting for test results or "the numbers."

We tell patients that their medical team fixates on test results so they don't have to. It's our job to gather all relevant information including complex measures like follicle counts, hormones levels, eggs retrieved, fertilized eggs, and all kinds of other mind-numbing numbers.

Probably the world's worst waiting room is the one you occupy while awaiting the results of tests related to your fertility, especially pregnancy tests. Again, there are ways to manage this. To counter the stress and fend off negativity, try repeating positive affirmations while you wait.

We encourage patients to repeat their favorite affirmations at least twice a day while also trying to find purposeful actions to take or upbeat things to do, whether it's redecorating the house, going to lunch with fun friends, or watching a comedy—preferably one that doesn't involve making babies.

Affirmations can be as simple as these:

- I will focus on the positive aspects of the situation.

- This will lead to something good no matter what happens.

- I will make the best of this situation.

- There are so many good things in my life. I will stay focused on them.

- I am here to take control of my life and find solutions.

- I will emerge from this stronger, wiser, and determined to make the most of my opportunities.

❀ DAMSELS IN DE-STRESS

The best approach for many patients is to combine mental approaches with physical actions. Once you have become aware of the benefits of controlling the thoughts and feelings swirling around in your brain, we encourage you to also consider the benefits of body awareness and methods for reducing stress physically.

This isn't simply a feel-good solution. The physical responses to stress can inhibit your ability to become pregnant, so being fit and relaxed and aware of your posture, heartbeat, breathing, and muscle tenseness can help more than you might think.

If an oncoming car suddenly veers at you, or a gun goes off as you walk in the woods, your body immediately goes into survival mode, preparing you for fight or flight. This is an automatic response to stress and it causes measurable changes in your body related to the release of adrenaline and cortisol. Your heart rate increases, you take short, shallow breaths, and your heart goes on turbo charge, pumping blood to the body's largest muscles in your arms and legs. Your mental awareness also sharpens.

In this highly aroused state you tend to perceive everything in your environment as a possible threat. If you happen to be dealing with infertility, this means you may overreact to threats that are not real, but only perceived, including spouses who just want to talk about "normal" things, nurses who insist that you follow the prescribed schedule, and the lady in the store who is exasperated with her children and has no idea that you would gladly trade places with her.

While this hyper-aroused state is a proven and useful survival mechanism, it is beneficial only as a short-term experience. Your body cannot tolerate long periods of rapid heartbeats, short and shallow breathing, or heightened senses. Unfortunately, fertility treatments often go on for several months, if not longer, so to maintain your health and to give yourself the best chance of getting pregnant, you have to find ways to avoid the fight or flight mode. If you can't relieve the stress on your body, you will be at risk for stress related digestive problems, high blood pressure, insomnia, obesity, fatigue, drug and alcohol abuse, and depression—as well as continued infertility.

❀ THE RELAXATION RESPONSE

We help patients counteract chronic stress by teaching them how to use the relaxation response, which is a state of rest and relaxation that produces feelings of calm and serenity. This helps decrease your heart and respiration rates, lowers your stress hormones, and reduces muscle tension.

There are many ways to learn how to elicit the relaxation response, but some of most effective techniques are:

- Meditation

- Yoga

- Progressive Relaxation

- Guided Imagery

- Breathing Techniques

Let's take a look at each of these methods and explore how they can benefit you.

Meditation

While meditation has long been practiced in Eastern nations, it has only become widely practiced in the U.S. in the last few decades. It is no longer restricted to monks, rock stars, and Californians. Meditation has gone mainstream. Even many leading corporations, including Procter & Gamble,

Prentice Hall Publishing, Nike, McKinsey & Co., AOL Time Warner, Apple, Google, and HBO offer their employees meditation training and facilities.

There are several types of meditation that can elicit the relaxation response, and most communities now offer classes or meditation practices through their schools or recreation departments. There are also many CDs and DVDs available to help you learn basic meditation techniques.

Most forms of meditation are very down to earth. First, you need to find a quiet place where you won't be disturbed or distracted. To begin, get in a comfortable position, close your eyes and focus on either your breathing or on repeating a key word or phrase that helps clear your mind and relax.

You don't have to chant a mantra, but feel free do to that if it works for you. Simply telling yourself to "relax" or to be at peace can help you accomplish that goal. Don't worry if you have trouble entering into a meditative state at first. Your mind will naturally wander, but as you become aware of this, you can gently bring your awareness back to either your breathing or your word. In fact, meditation trainers say that the simple practice of realizing that your mind has wandered and bringing it back to a meditative state will strengthen your ability to meditate each and every time you do it.

As you master meditation and the ability to stay in that calm and serene state, you will find your heart rate decreasing and your breathing slow down. Some people feel like a gentle veil drops around them, enveloping them in relaxation. When you feel this happening, enjoy the moment and know that you have succeeded in reducing stress.

It may be difficult, at first, to meditate for longer than five or ten minutes at a time, but you will find that the more you do it, the longer you are able to sustain the meditative state.

We advise practicing meditation at least once a day for thirty minutes or more once you're able to sustain it. Like exercise, it will get easier with practice, and the more often you do it, the more benefits you will feel.

Many people feel self-conscious or can't relax at first. If you find yourself fighting it, you might try listening to meditative music or to an audio recording designed to help you meditate. As with everything else these days, there are smart phone apps designed for meditation too. Ferring Pharmaceutical has a comprehensive app called FertiCalm that you can download for free. Some popular choices have catchy names such as Mindful IVF Buddhify, Omvana, and Headspace. There are many others and most are offered at no cost in their basic forms.

The relaxation response is just as natural and automatic as the fight or flight response. Anyone can learn to elicit it. However, it can take a concerted effort and commitment to get into the daily habit of doing it—just as it takes a daily effort to exercise. You might find it easier to join a meditation class, or to do it with a friend who has more experience.

Like many of our patients, Emily found that the more she practiced it, the better it worked for her.

Emily did most of her meditating at home after taking classes at her church. Many faiths use meditation to reach a spiritual state, and as a form of prayer. At home, Emily went to the room set aside to be the baby's room. She meditated there for a half hour or so early in the morning. She'd sit in a chair and use CDs she'd recorded of herself talking through steps to help her control her breathing, to visualize positive outcomes, and to reach a meditative state.

"I would imagine myself on the beach relaxing and then I would visualize myself pregnant," she said. "I had a whole script that I would run through in my head. My goal was to clear out negative energy and to see myself where I wanted to be, as if it had already happened."

Yoga

Even before she wanted to start a family, Emily had practiced yoga, another form of stress reduction that was once considered "out there," but is now as popular as yogurt across the United States. There are many forms of yoga available nowadays and most communities offer classes through public recreation centers, YMCAs and YWCAs as well as privately owned yoga studios or classes offered to members at local fitness clubs.

Most forms of yoga involve holding poses while controlling your breathing. Some forms of yoga also emphasize physical exertion, while others focus on meditation. Yoga is meant to strengthen and soothe both the body and the soul. Practicing yoga helps you become more aware of the tension points in your body so that you can work to release tension, alleviating soreness, aches, pains, as well as stress.

We do advise patients who are taking fertility drugs to stay away from the more rigorous types of yoga including "hot yoga" or any other highly strenuous exercises that might cause harm to enlarged ovaries. Gentler forms, including "restorative yoga," are generally okay for our patients undergoing treatments. In fact, our clinic offers free downloads to patients of a DVD called Restorative Yoga for Fertility. Many of our patients have found it very helpful in promoting healthy relaxation during treatments.

<u>Progressive Relaxation</u>

Progressive Relaxation is similar to yoga, though not as strenuous. This technique that has been around since the 1920s to help reduce anxiety. There are a couple forms of progressive relaxation. One involves becoming aware of the tension in each part of your body and intentionally increasing it prior to letting the tension go. The other involves simple awareness of the tension and then letting that tension go.

To help our patients practice this technique, we offer CDs and MP3s entitled "Guided Relaxations for Mind/ Body Wellness," with tracks that offer progressive relaxation methods for daily use, as well as for use in conjunction with your medicated cycle. We encourage patients to use these methods throughout treatment and even when they are having an embryo transfer.

We also offer these steps to promote progressive relaxation:

- Lie down in a comfortable place where you won't be disturbed for ten to twenty minutes.

- Close your eyes and take a few deep, gentle breaths.

- Focus on your face and especially the area around your eyes. Notice any tension you are holding there. As you breathe in notice where you are tense, and as you breathe out, let go of that tension.

- Breathe in relaxation.

- Breathe out all the tension.

- Now move your awareness to the muscles around your mouth and notice where you are holding tension in that area.

- Breathe in relaxation to that area.

- Breathe out all the tension.

- Move your awareness to your jaw. Notice where you are tense.

- Breathe in relaxation to your jaw muscles.

- Breathe out the tension.

- Become aware of your neck. Observe where you are holding tension.

- Breathe in relaxation.

- Breathe out all the tension. (Notice the difference in how your body feels.)

- Move your attention to your shoulders. Notice where you are tense.

- Breathe in deep relaxation.

- Breathe out the tension.

- Become aware now of your upper arms and the tension you may be holding there.

- Breathe in relaxation.

- Breathe out the tension.

- Turn your attention to your lower arms and hands. Note where you are tense.

- Breathe in relaxation.

- Breathe out all tension.

- Concentrate now on your stomach and pelvis. Notice where you are tense.

- Breathe in deep relaxation.

- Breathe out the tension.

- Focus on your thigh muscles and buttocks. Note where you are feeling tense.

- Breathe in relaxation.

- Breathe out the tension.

- Become aware of your calf muscles. Notice any tightness or tension.

- Breathe in relaxation.

- Breathe out the tightness.

- Focus on your feet and especially your toes. Note where you are tense.

- Breathe in relaxation.

- Breathe out all the tension.

Now, for the grand finale!

- Breathe in relaxation to your whole body.

- Breathe out any remaining tension.

Finally, take a few minutes to enjoy feelings of peace and relaxation.

Guided Imagery

Guided imagery is a powerful technique that uses your imagination to create a more relaxed and focused state. Athletes use this technique to visualize and improve their performance. It is a form of visualization that involves concentrating on images that can evoke profound feelings of peace, confidence, and relaxation.

You will often hear quarterbacks talk about visualizing plays before a game, or downhill skiers visualizing themselves making runs before a competition. You may even have unconsciously done this yourself by mentally rehearsing a dinner meeting, or a speech, or walking through a presentation in your mind.

You can create your own images, of course, but it can be helpful to listen to someone else describing these images, especially if you are a rookie at this. We provide our patients with a CD titled *Guided Relaxations for Mind/Body Wellness*. The third track on that CD is called "Relaxing and Reclaiming the Body's Power and Strength." It offers some very effective instructions for imagining a healthy, strong, fertile body.

If you prefer to do your own guided imagery, we offer these steps:

- Sit or lie down in a comfortable place where you will not be disturbed of ten to fifteen minutes.

- Think of a place where you love to be. It may be a beach or place in the woods, or lovely house or room where you were happy and completely content. It might be a place you visited recently, or somewhere in your past.

- Remember what that place felt like—the warmth of the sun, the feel of the sand, any sounds you remember that you particularly love.

- Bask in the comfort of your favorite location. Let yourself return mentally, reminding yourself why you love it so much.

- Stay there as long as you like in your mind as you relax.

- When you are ready, return to where you are right now, and take that feeling of pleasure and contentment with you.

- Anytime you are feeling anxious, angry, or stressed, you can return to this place.

Emily was going through the IVF treatments when she used our CDs and she found them quite useful in helping her relax.

"I played the CDs each night and it was so relaxing, I'd fall asleep quite often in the middle of it, which is a good thing because the message still registers in your brain as you sleep," she said. "It's all about thinking positively, so doing that and meditation helped calm me down too."

Breathing Techniques

The way you breathe profoundly affects how you feel. If you don't believe this simple fact, try *not* breathing for the next ten minutes. See, that doesn't feel good does it?

When you are anxious you tend to unconsciously hold your breath, or take shallow ineffective breathes. This can make you feel dizzy or faint, which increases your anxiety and your worrisome thoughts. If you aren't fully aware of what you are doing to yourself, the consequences can be rather unsettling, even dramatic. You might faint or feel as though you are having some sort of panic attack.

We want you to be aware of what is going on in your body and how it affects you physically and emotionally. You can do this by paying attention to your breathing and its impact on your mood. Simply checking in throughout the day to monitor your breathing can be helpful. Am I feeling stressed? What's my breathing rate? If I slow it down, will I feel better?

Beyond that, the following techniques can also help. You may be surprised at how quickly you can decrease your anxiety by taking even a few minutes to do the following exercises:

Abdominal Breathing

- Place your hand on your abdomen.
- As you breath in, inhale deeply enough so that you can feel your hand lifting slightly.
- As you exhale, feel your hand dropping as you abdomen constricts.
- Repeat five to ten times.

Timed Breathing

- Take a deep breath, to the slow count of 1, 2, 3.
- Exhale slowly to the count of 1, 2, 3, 4.
- Repeat five to ten times.

Alternate Nose and Mouth Breathing

- Inhale deeply through your nostrils.

- Exhale through your mouth, allowing your mouth to be loose, making a blowing sound.

- Repeat five to ten times or until you feel the relaxation response start kicking in.

❀ *FINDING WHAT WORKS FOR YOU*

There are many other forms of relaxation and stress reduction that our patients have used successfully. Emily is among those open to alternative methods. She went gluten-free, jogged ,and walked each day, and also tried acupuncture— as have many of our patients—as well as hypnotherapy and the eastern method of massage known as *Reiki,* which some believe activates the body's natural healing processes.

Emily can't say specifically that any one of these methods worked for her. All she can say is that, in the end, she was able to have a child after years of struggling. She notes, however, that while having a child is a beautiful blessing, most women will still benefit from stress-reducing tools and techniques.

"I will never forget what happened, but I feel healthier now. I no longer have the obsessive thoughts. However, life is not perfect.

"Life always has its challenges and the trauma of infertility can take a while to deal with. On top of that, raising a baby is not always an easy walk in the park. Then there's the fact that some of us have a sort of survivor's guilt. I have friends who still can't get pregnant and I feel guilty when I am with them. I also have friends who spent one hundred thousand dollars or more to get pregnant and they can be resentful toward women who have children easily and naturally."

Emily's point is a good one. Even if you achieve pregnancy and have children, you may still have to work to fend off negative thoughts and debilitating emotions. We encourage you to find what works best for you, and to understand, that through struggle, we can gain strength, wisdom, and empathy for our fellow human beings.

"I thought I was a good person who established good rapport with my clients, but my goodness, this has, hands-down, made me a better social worker and a kinder person," she said. "I've learned to care for others more and I am more alert to the fact that if people are acting certain ways, they probably are going through something difficult like this and their emotions are raw. I'm more aware, more perceptive, and more empathetic."

Many of our patients feel the same way, and we believe that is a very healthy mentality. We learn a great deal about ourselves through adversity, and by rejecting negativity, managing our emotions, and taking care of our minds as well as our bodies, we can become better, stronger, and more caring human beings.

You never forget your struggles, but once your infertility is resolved, your obsessive worry and fear of failure will likely lessen considerably. There are even significant benefits to dealing with adversity in that it increases ones sensitivity to others and affirms our sense of personal strength, making us better, stronger, and more caring human beings.

CHAPTER TWELVE

THE MANY BENEFITS OF FERTILITY ASSESSMENTS

BY DOCTOR ROB GUSTOFSON

Around the time she turned thirty-five, Rochelle, an IT professional in Denver, saw her twin sister and an older sister go through miscarriages. Even though both sisters already had successful pregnancies and healthy children, their miscarriages gave Rochelle cause for concern. She decided to do something she'd been putting off: getting a fertility assessment so she could know if there were any challenges she might face in having children one day.

Rochelle was single and although she was an avid snowboarder and in great physical condition, she knew her most fertile years were behind her. Her main concern, she says, was "fear of the unknown."

"I have been extremely healthy my entire life, but there was just that tinge of doubt that there might be something wrong, something I didn't know about, or an abnormality I had no idea about," she says. "Getting a baseline of my (reproductive) health was important to me. Any fear I had was outweighed by the opportunity to know where I stood.

Even if they found an abnormality, at least there would be information available so I could come up with a plan and know what I was dealing with."

We recommend that women over twenty-five have fertility assessments for much the same reasons. If a fertility issue is detected early, it allows the patient more time and options to take steps to preserve fertility and plan for a future pregnancy. All too often, patients find out at a later age there is a fertility problem and that makes it more difficult for them to get treatments so they can have children.

This is especially a concern for women around Rochelle's age. If they wait to have children, any further delays to correct fertility challenges will likely make it more difficult for them to conceive because of the normal effects of aging alone.

With the advancements in egg freezing over the last five years, more women can now benefit from fertility assessments that let them know the status of their egg quality and whether they have other challenges awaiting them. Women whose eggs are found to be healthy and viable might feel more comfortable waiting to start their families depending on their age. Women whose eggs are of poor quantity might consider preserving their fertility by freezing eggs or embryos before further compromise occurs because of age or health issues.

❀ AM I FERTILE?

The basic question Rochelle and most other young women have when they come to us is, "Am I fertile?" Such a simple question begs for a simple yes or no answer. With advances in science, technology, and innovation, it should be a quick answer yielding black and white results. The reality is, unless you are a woman who is currently pregnant or a man who has fathered that pregnancy, the answer often is neither yes nor no, but lies in the ambiguous area between.

The Gray Zone is a world of statistical probabilities involving many personal historical factors, analytic testing, and keen medical judgment. It is far from simple for women or men.

Let's start with a basic fact: a woman under thirty-five has an 85 percent chance of conceiving a pregnancy within the first twelve months of trying, provided she is without medical problems, has regular menstrual cycles, and is having intercourse around the time of ovulation with a male partner who is also without medical problems.

A couple would be deemed infertile technically if the female partner under thirty-five has failed to conceive after a year of trying. For women thirty-five and older, trying six months without success would be adequate cause for seeking help because technically this is considered infertility.

A fertility assessment provides a statistical estimate of an individual's chance to conceive prior to trying. But how do you evaluate a woman or man who has not yet attempted pregnancy? If they haven't tried to conceive, how can they be infertile? Can you deem someone fertile or infertile only? Can someone wait a few years to try? Should that person do something to preserve fertility? Does an abnormal test mean you are infertile? Could you conceive without help if you've tested abnormally? These are among many of the questions that may arise.

❀ THE TICKING CLOCK YOUR GRANDMOTHER TALKED ABOUT

For women, the most important measure for predicting the chance of pregnancy is age. Your age is the battery that powers the biological clock and causes the clock ticking to become louder and louder.

The hard truth is that increasing female age results in decreasing pregnancy rates and increasing miscarriage rates. This means it becomes harder to conceive and the likelihood of a successful delivery decreases due to decreasing egg health. This doesn't mean a woman can't conceive or that she is not fertile, it means the chances are decreasing as she ages.

Again, that's not to say all older women are infertile. There are plenty of anecdotal stories of women in their late forties who have conceived without any assistance. Even women who were deemed infertile have surpassed the odds and managed to conceive healthy, happy children.

When you hear of a woman in her forties who has had a healthy baby, the question to ask is "How many women in the same category tried and didn't conceive compared to the one that did?" The chances of conceiving are never zero based on age alone, but the odds do begin decreasing exponentially after age thirty-five.

❀ A FERTILITY ASSESSMENT

Women are born with millions of eggs (five to seven million); however, the number of eggs decreases progressively from birth (one to two million) to puberty (five hundred thousand) and then menopause (one thousand). The eggs age at different rates among women regardless of their health status.

In general, the quality of a woman's eggs decreases progressively from puberty. As Rochelle learned, more rapid decreases in quality begin at the age of thirty-five with increased difficulty conceiving naturally in the late thirties and forties. The one thing that was shocking for her, she says, was learning about the rapid decline in pregnancy success rates from thirty-five to forty and beyond.

"When my doctor told me that, it was the 'Aha!' moment for me," Rochelle says. "The biggest surprise for me was that between twenty-five and thirty-four the chances of getting pregnant are virtually the same each year, but after that it starts dropping fast and after the age of forty, it drops significantly."

Unfortunately for some women, the quality of their eggs decline as early as their twenties. Over the years one of the common frustrations women have voiced to our team is that they wish they had known their eggs were aging at a faster rate than normal.

We believe that with the advancements in egg freezing over the last five years, more women would benefit from knowing the status of their egg quality as early as feasible. Women whose eggs are still viable can consider delaying starting their families depending on their age. Those whose eggs have a decrease in quality or quantity may consider preserving their fertility by freezing eggs or embryos before further compromise occurs based on age or other medical factors.

There are also women whose mothers or siblings have experienced infertility and were diagnosed with decreased egg quality or quantity at a young age. These women would also benefit greatly from understanding their egg health early in their lives.

❀ *ASSESSING YOUR FERTILITY*

CCRM offers women of child-bearing age an opportunity to determine the current status of their ovarian reserve through our Fertility Assessment Program. In order to qualify, women must be under the age of forty, not planning to conceive within the next six months, and desire information about her current fertility status.

Women who are on birth control pill are asked to discontinue the pill for one menstrual cycle before the testing is completed. You can meet with the physician first before you stop the pill if you are uncomfortable stopping the birth control pills prior to your first consult. In addition, you should consider an alternative, non-hormonal form of contraception if you are sexually active, such as condoms, to prevent pregnancy.

The following services are included in our fertility assessment program:

- new patient consults with one of our staff physicians

- 3D Baseline Ultrasound to evaluate the uterus and the number of eggs potentially stimulating during that menstrual cycle while also assessing both quality and quantity of eggs

- day three hormones, including FSH, LH, Estradiol Lab Work, Anti-Mullerian Hormone (AMH) Lab Work

- regroup with your physician to review test results

❀ *PRETEST PROBABILITIES*

The first step to answering the question "Am I fertile?" begins with looking at your pre-test probability, which is the likelihood that a patient is infertile based on personal history. If a woman's mother and female siblings have had fertility challenges, then there is more likelihood that she will have them. Smoking, heavy drinking, using painkillers, or being severely overweight or underweight may impact her fertility as well.

A healthy woman who is twenty-eight years old, and menstruating regularly with no health concerns is likely to be fertile based on her age alone. Diagnostic testing will help to reassure her that she has good egg quality and, therefore, she is fertile at the time of testing. On the opposite end, a woman who is forty-one years old and equally healthy still has a much lower probability of fertility based on her age alone. Diagnostic testing will help to determine if she has fertility chances equal to or worse than what her age would predict.

❀ *DIAGNOSTIC TESTING*

For decades, scientists and clinicians have worked to find the "one" test that would predict who can or cannot conceive. For now, there is no single test that does it all. Instead, we do a series of tests that provides a composite view of the egg health at the present time. It does not predict how rapidly your eggs might decline in the future. Even if you test with normal fertility at one age, there will always be a decline as you grow older, but we can't accurately estimate the slope of the curve.

The three tests measure levels of:

- FSH/Estradiol (which stimulates the ovaries to produce eggs)/estradiol (the primary female sex hormone)

- Anti-mullerian hormone, a hormone produced by the cells that support the dormant pool of eggs in the ovaries. (Therefore, a higher level of AMH suggests a larger pool of eggs or a greater ovarian reserve.)

- Resting/antral follicle count, an ultrasound test to measure your future egg supply also known as your ovarian reserve.

There are many other tests that have been considered but they are not consistently used in clinical practice.

Let's look at each of these three tests done in our fertility assessments.

Test 1: Follicle stimulating hormone

FSH, or follicle stimulating hormone, is the way that the pituitary gland in the brain communicates with the ovaries. The FSH level is assessed by bloodwork early in the menstrual cycle, typically on day two or three. It signals the ovaries that it is time to make an egg each month. As the eggs decrease in quality, they are more resistant to FSH and it takes more and more FSH to keep the cycle progressing.

Therefore, the higher the FSH value, the lower the quality of the eggs. A normal FSH value is less than 10 uIU/mL when assessed early in the cycle. A normal FSH is good but gives limited information. The most helpful value, although most concerning, is one that is elevated.

FSH will vary from one menstrual cycle to the next. Researchers have tried to use a lower FSH value to predict when it is best to stimulate for pregnancy during in vitro fertilization cycles. Unfortunately, this has not worked well as it appears a woman's eggs are only as good as the worst

FSH value she has obtained. For example, a woman has an FSH value of 5.6 uIU/mL in one cycle and then it increases to 15.1 uIU/mL in the next cycle. Her chances of pregnancy are predicated on the 15.1 value and not the 5.6 value. This fact has been demonstrated in numerous studies and persists even today.

Estradiol is the companion to FSH and is the ovaries' method to communicate with the brain. Estradiol levels increase as an ovarian follicle becomes stimulated and is increasing in size. For evaluation and to ensure the validity of the FSH value, the estradiol level should be under 50 pg/ml. If it is elevated above 50 pg/ml early in the cycle, the ovaries are responding early and the quality is potentially decreased.

Test 2: Anti-Mullerian Hormone

The most recent addition to the arsenal of ovarian reserve testing is anti-mullerian hormone (AMH). AMH is a glycoprotein released by the granulosa cells of early developing eggs that help protect and mature the eggs. Over a lifetime, as the number of eggs decreases, the granulosa cells decrease and AMH decreases. AMH gives an estimate of the total number of eggs that are remaining in the ovaries.

Overall, studies have demonstrated this value provides an insight to the number of eggs that can be retrieved in in vitro fertilization and quality of the embryos; however, results have varied in regards to pregnancy rates. There is no level below which pregnancies have not occurred.

Although high levels are associated with more eggs in the ovaries, it can be associated with other conditions like polycystic ovarian syndrome. AMH can be drawn at any point in a menstrual cycle as well as while on birth control pills. Birth control pills will lower the AMH level but can be reassuring if it is normal.

Test 3: Resting/antral follicle count

The final parameter that has been utilized to evaluate the quality and quantity of eggs is the resting/antral follicle count (RFC). The RFC is assessed by ultrasound in the early follicular portion of the menstrual cycle, usually between days three to twelve of menses. It is performed by counting the total number of follicles between four and nine millimeters in both ovaries. A normal value for RFC is greater than twelve. RFC will have intrinsic variation from month to month and will vary based on when it is performed in the cycle as well as based on the ultrasound technician.

❋ THE EXPERIENCE

On her first visit for her fertility assessment, Rochelle had her ultrasound and blood drawn. She returned for hormone testing on another day. The only unpleasant part was the ultrasound test. "Having a camera down there was not pleasant, but as a woman you are used to being poked and prodded on your annual physical, so I've had worse," she says.

"I was pleasantly surprised that my doctor felt more like a partner in the process than a typical physician-patient relationship when I went back for the test results and assessment review," Rochelle says.

She was pleased that there was no pressure from her doctor or staff members to make a decision on what to do right then and there. "She gave me the numbers and told me what studies have shown as far as fertility rates as you age,"

Rochelle notes. "She said whenever I was ready to make a decision on the next step she would be there for me."

Rochelle's numbers were good as far as the condition of her eggs. They did find a cyst on her ovary, but it was not considered to be a problem or a threat.

"That was a little disconcerting, but they said they would keep an eye on it to make sure it was benign," Rochelle says.

While she found the medical terminology involved to be "pretty intimidating," Rochelle thought her doctor did a very good job of explaining each step and procedure to her in terms she could understand. "Mostly as a patient, you just want to hear you are in the normal range and beyond that you shut out all the other stuff while you are thinking, "'Yay, I'm okay.'"

❀ EVALUATING THE EVALUATION

Once you have completed testing, how is it interpreted? Which is the most important value? How do you estimate likelihood of success?

When considering the testing, the four parameters are evaluated in order of importance:

- Age

- Resting follicle count

- Anti-mullerian hormone

- FSH

The normal values do not change based on age. Consider the following example: A woman's results show her with an FSH of 10.7 mIU/ml, estradiol of 35 pg/ml, AMH 1.2 ng/dl, and resting follicle count of nine. These values show she has diminished ovarian reserve and pregnancy rates will be less than predicted by age. These values are of concern regardless of age, however, less concerning if she is a twenty-eight-year-old compared to a thirty-nine-year-old. A twenty-eight-year-old will still have good chances to conceive with special treatments whereas a thirty-nine-year-old should consider in vitro fertilization as the first line of treatment.

Each value imparts information about chances of successful conception. Rough estimates of pregnancy can be made by the lab testing. For example, an RFC under five has been associated with a 50 percent reduction in pregnancy compared to age. Similarly, an FSH between 10-14 mIU/mL has been shown to reduce pregnancy rates by one-third to one-half age related comparisons.

Take for example, a thirty-nine-year-old woman with an FSH of 21.1 mIU/mL, AMH of 0.1 ng/ml and RFC of two. Her lab tests are all congruent with each other. Her assigned probability of conceiving each month—even if using in vitro fertilization--would be estimated to be less than 1 percent due to severe diminished ovarian reserve. Again, this does not mean she cannot conceive, only that the likelihood is very low. The combination of these parameters assists clinicians in assigning a relative chance of success to conceive.

❁ DIFFERENT SEXES HAVE DIFFERENT CLOCKS

Women and men age differently with regards to reproduction. Women, who bear the brunt of reproduction, have a rapid decline in fertility with age. Men may reproduce, however, throughout their lifetime—just ask actor Tony Randall. The persnickety half of *The Odd Couple* television show in the 1970s married a twenty-five-year-old woman when he was seventy-five, and had two children with her. The last was conceived when he was seventy-eight-years-old. Take that, Oscar!

The male biological clock doesn't seem to stop ticking at any particular age. Sperm parameters, like volume of ejaculate and concentration of sperm, decrease over time with a mild decrease in fertility. Increasing research is demonstrating that there are, in fact, age changes for men: increased risk of chromosomal abnormalities, associations with psychiatric disorders like schizophrenia and bipolar disorder, autism spectrum disorders, and other rare genetic disorders like Crouzan's Syndrome. These increase slower and in lower frequencies than the effects of age with women.

Assessing male fertility is a simple process that requires complex evaluation. This is done with a semen analysis. The appropriate collection of sperm is performed after two to five days of abstinence from sex and analyzed within thirty minutes of production. The specimen is evaluated for many parameters: volume of the ejaculate, pH (acidity or basic nature), viscosity, concentration or amount of sperm per milliliter of fluid, percent of sperm moving, speed of movement, and morphology or shape of the sperm.

One test does not completely predict the ability to conceive. Similar to female testing, the analysis will vary with each collection as well as any past or present medical conditions: illness, substance abuse, and excessive heat. Male sperm production takes about seventy-two days from creation to ejaculation.

Similar to AMH, there is no sperm count below which a pregnancy may not occur, except when no sperm are present at all. Again, it is about the probability of pregnancy based on studied parameters. Unlike women, however, a male's age is not the most important factor. Concentration, motility, and morphology of sperm gives the best estimate of pregnancy.

In a landmark study published in the *New England Journal of Medicine* in 2001, fertile parameters were reviewed. These parameters vary from the typical accepted lab values for the World Health Organization. Listed below are the generally accepted parameters of sperm, below which are associated with subfertility:

- Concentration—13.5 million/milliliter

- Motility—32 percent movement of sperm in a specimen

- Morphology—9 percent normal shape as assessed by Kruger strict criteria

The most important parameter for sperm is morphology. When sperm have a morphology below 9 percent, the risk of infertility increases by 2.9 times. When there is a trio of abnormalities including concentration, motility and morphology, risk of infertility increases to 15.8 times.

Sperm evaluation continues to evolve. Other examinations of sperm include antisperm antibodies, DNA fragmentation through various assays, culture of semen for infections, binding studies, penetration assays, and several other methods to assess fertility. These assays are used pending the clinical scenario and are not usually considered routine for a simple assessment of fertility potential. Most are used in the male or couple deemed infertile to assist with planned care.

Men who have abnormal assays should be evaluated by a urologist to find potential reversible causes. In the event that no reversible cause can be found, men can consider sperm banking through a cryobank to ensure that sperm is available for future attempts for pregnancy.

● *BENEFITS OF THE FERTILITY ASSESSMENT*

Based on her results and the information she received from her doctor on the rate of egg declines in her age range, Rochelle is considering freezing her eggs within the next year.

Rochelle had been concerned about the cost because she'd initially heard that the egg freezing would cost seven thousand dollars to eleven thousand dollars or more depending on the clinic. The typical three hundred dollars per year egg storage cost is another factor that she will have to consider, but Rochelle did not find the cost daunting.

With advancements in fertility care, women now have an array of choices for action after a fertility assessment. One consideration includes being reassured by the evaluation and doing nothing. Keep in mind, the evaluation is limited to that moment and cannot predict how fast a decline may occur.

Others who are considering pregnancy "sometime" may opt to move up their timeline if there appears to be a decline in chances. This may include options with a male partner or utilization of donor sperm.

Women like Rochelle who are not currently in a relationship can consider egg freezing to preserve fertility at their current age and status. This leaves the option of using a partner's sperm or donor sperm in the future. An increasingly popular option for women is to freeze their eggs for later fertilization.

"Doing the assessment helped me decide to do the egg freezing, especially because deterioration rates are so alarming," she said. "My doctor said even though I am healthy and likely to get pregnant, the older I get, the harder it will be. So, I will see how I will feel toward the end of this year and early next year, and then I will probably start the process by next year."

Freezing her eggs will give Rochelle options for the future. For now, she says she will probably use the eggs only if she marries because she doesn't want to raise a child on her own. Fertilizing the egg and freezing embryos also gives women like her another chance to maintain pregnancy rates at their current age.

The most frequently asked question among patients considering preserving their fertility is, "Am I fertile?"

My answer is a simple I don't know. This is obviously frustrating to her because she is here to answer the most important question on her mind.

After doing a fertility assessment, the clinician can provide her with an idea of her chances and options to maintain her fertility for the future. There is no crystal ball to predict her future pregnancy chances and, therefore, she can consider fertility preservation. Science continues to evolve and create options for fertility assessment, care, and preservation. Until more refined testing is created, the best option for fertility is to be proactive at as young an age as possible.

Kate said her good results comforted her, but the information she received has also motivated her to protect her fertility.

"I would say if you want to have kids, at least do the assessment," she says. "It gives you the peace of mind. It is one of those steps that as an older female, you need to take in order to understand where you are starting from. It will eliminate a lot of frustration if you can't get pregnant."

CHAPTER THIRTEEN

RAPID ADVANCEMENTS IN REPRODUCTIVE SCIENCE

BY DOCTOR WILLIAM SCHOOLCRAFT, M.D.

When I was just beginning my medical education, the idea of growing a human embryo in a test tube was almost unimaginable. Then, just one year before I earned my medical degree, the first "test tube baby" was born. (She now has three children of her own.) You might say this development inspired my life's work. Across our network of 11 centers in the past thirty years, CCRM has helped more than fifty thousand people become parents.

We also have worked on the cutting edge to develop stimulation protocols that enabled women to produce multiple eggs, and we have played a role in the creation of methods for injecting a single sperm into an egg to achieve fertilization, helping men who might never have had a chance at fatherhood.

Other major advancements we've helped develop include the creation of a laboratory culture media that allows us to culture embryos on an extended basis to the more advanced blastocyst stage, which gives women a much better chance of becoming pregnant. Prior to that, it was widely believed that embryos could not be cultured past the eight-cell stage. As a result of this advancement, the technique of embryo biopsy, and testing all twenty-three pairs of chromosomes at the blastocyst stage also became a reality.

Again, this was something we could not have imagined even fifteen years ago—and it seems that the timetable for advances keeps moving faster. Just a year or so before I began working on this book, doctors in Sweden became the first to successfully transplant a donated uterus to a woman who had been born without one. The twenty-six-year-old woman received a donated womb from a friend who was in her sixties and had gone through menopause seven years earlier. They had to use drugs to suppress the recipient's immune system so her body wouldn't reject the transplanted uterus, but in the end it was successful.

The woman and her partner had frozen embryos through IVF, and a year after the transplant, their doctors transferred one frozen embryo. There were some complications. The baby was delivered early, at thirty-two weeks, after the mother developed pre-eclampsia, but news reports said they were doing well.

In February 2016, doctors at Cleveland Clinic performed the first uterine transplant in the United States. The patient, twenty-six-year-old Kate McFarland has three adopted children with her husband, but she had always wanted to carry and give birth to her own child. She received the uterus of a thirty-year-old woman who had died suddenly. Sadly, just two weeks after the transplant, doctors had to remove the uterus because of an infection caused by a common "yeast infection" fungus found in women's reproductive systems.

The Cleveland Clinic case shows that there are still many challenges that must be overcome before uterine transplants will become widely accepted. Still, this is another remarkable advance in reproductive science and could well be a major benefit for women who have birth defects affecting the womb, or for those who have undergone cancer treatments.

❀ *IN-VITRO ACTIVATION HOLDS GREAT PROMISE*

There are so many amazing developments like this that it's hard to keep up with all of them these days, but we certainly do our best—and often, we are involved in the research and development. In 2015, I went to Japan to check for myself on another major new advance in reproductive science. This technique, which we hope to help refine for our patients, may allow women who have gone through menopause to have children—something we'd thought was not possible in the past.

This process is called in vitro activation or IVA. First developed at Stanford University, it was tested in Japan. Their 2013 study involved thirty-seven women. The Japanese researchers collected mature eggs from six of the women. Two of the women had successful pregnancies and now have children. Since then, five more pregnancies have been reported. This outcome received a great deal of attention in our field because of it offers hope to so many women previously thought to be unable to have children.

The procedure begins when all or part of a woman's ovary is removed from her body. Then, the cortical tissue is divided up into small segments, and then implanted back near her fallopian tubes. The woman is then given follicle-stimulating hormones to increase her chances of pregnancy. The thought is that this would allow primordial or small follicles in the tissue to be awakened so they would develop into mature follicles and, hopefully, eggs.

The theory is that switching on dormant follicles may allow eggs (that never would be allowed to ovulate otherwise) to be rescued from the ovarian tissue. This work is in its preliminary stages as well, but we are hopeful that significant progress can be made in this revolutionary concept over the next few years.

The interesting thing is that women are born with a million eggs and even when they're going into menopause, there are a few thousand eggs left in the ovaries. Those eggs are in an immature state where they can't grow and ovulate. Those last few thousand eggs won't come out and they won't ovulate, so the woman's ovaries act like they're not working, but if you look in the microscope, you can see they are there.

This is an aggressive treatment and still considered experimental, but I am excited about the positive outcomes achieved in Japan. For those patients whose only previous option was donor eggs, this is an alternative that allows them to instead have an opportunity to have a baby with their own genetic makeup.

Of course, we've recently moved into an era when younger women can freeze their eggs long before they've experienced menopause. Better methods of freezing embryos have been developed and this, in turn, has also resulted in greater success with egg freezing so that women can preserve their fertility until they are ready to have children.

This has been a boon for women who want to accomplish other things prior to starting their families. I have seen so many mind-boggling and unimaginable developments over the last thirty years of my career. I have no doubt we will see equally dramatic advances over the next decade or two. The future of fertility treatments is bright indeed.

❂ *ADVANCES IN EMBRYO SELECTION*

I fully expect also that there will be great advances in our ability to identify the healthiest and most desirable embryos for use in IVF treatments. The major drawback to IVF treatments for infertility in the past has been the risk of multiple births. While there are certainly some people who are glad to have twins or triplets, it is also true that the risks involved in those pregnancies are greater than with single births. In the past, IVF usually has involved the transfer of two or more embryos in an effort to optimize success, but of course that often results in multiple embryos and births as well as increased complications for both the babies and the mothers.

We've long believed that the perfect scenario would be to transfer one embryo and have 100 percent successful pregnancies and deliveries. We may never achieve the perfect level of success, but it is a goal we are still striving for. A key to reaching this goal is developing the ability to look at all of a patient's embryos and then accurately determine which of them offer the best probability for a successful pregnancy and delivery.

Science is moving us closer to that goal, thanks to our ability to use the more mature blastocyst stage embryos. As I mentioned earlier, our CCRM team played a significant role in this advancement based on our realization that the environment inside the womb changes as embryos develop. We came up with laboratory culture media more like that in the womb so we could nurture the embryo to the more mature blastocyst stage. In addition, we developed methods for chromosomal screening of the blastocyst embryo, to help find those most likely to lead to a successful pregnancy and delivery.

Still, roughly half the embryos screened in these fashions fail to implant. Thus, we must keep striving. Areas that we are working on at CCRM—along with other scientists around the world—involve OMICS technologies, which explore the roles, relationships and actions of the various types of molecules that make up the cells of an organism. Genomics, the study of genes and their function, is one example of an omics technology. We also can look at metabolomics, which is the metabolism of eggs and embryos and learn a lot about their physiology.

Our ability to look at embryo glucose uptake, lactate production, amino acid turnover, and other metabolic parameters has led to increased success of embryo implantation. At CCRM, we have pioneered methods for assessing proteins produced by embryos as markers of viability. This is part of a field known as proteomics.

Another advancement, time-lapse imaging of embryos, has gained traction in the past few years and as the technique is refined, will probably also serve as a marker to determine the single best embryo, or most viable embryo, for transfer. Finally, gene expression, or different genes that are turned on and turned off by embryos, as well as their surrounding support cells, such as cumulus cells, may give us valuable insights into the physiology of the most viable and successful embryos.

❁ SEEKING A GREATER UNDERSTANDING OF THE UTERINE ENVIRONMENT

The uterine environment is another area of great interest in the rapidly advancing field of infertility treatment. The major question is what makes a uterus receptive or unreceptive to an embryo. While we now have many high-tech ways to diagnose and examine the uterus for abnormalities, there still are many patients with a seemingly "normal" uterus that do not allow embryos to implant.

We are among those scientists working on this problem by assessing the environment in the uterine lining at the time of embryo implantation. Some groups have taken embryo biopsies and studied genes that are turned on or turned off at the time of implantation. Others have studied the secretions in the opening (lumen) of the cavity and examined them with proteomic technology for analyzing proteins and metabolomics technology that allows the study of the unique chemical fingerprints left by specific cellular processes.

At CCRM, we assessed these secretions immediately before transfer and we are using this data to determine the best day for transfer, whether the uterus is receptive, or whether we should leave the embryos frozen and wait for a different cycle when a more favorable outcome will occur. We believe our ability to do real-time monitoring of uterine receptivity will improve our ability to move closer to the goal of one embryo/one baby 100 percent of the time.

❋ OVERCOMING POOR EGG QUALITY

Perhaps our most vexing problem as this book is written is finding ways to help patients overcome poor quality eggs, or what we call diminished ovarian reserve. While we can try to treat these patients with our best technologies, if the eggs are not healthy, and they are few in number, we have no treatment to overcome those issues. The next Holy Grail in our field, then, would certainly be to take eggs from older women that are of suboptimal quality and make them better. I have no doubt that in the next decade there will be breakthroughs in solving this challenge.

Currently, the only successful option is to bypass the problem by using healthy eggs from a donor instead of the patient's unhealthy eggs. While egg donation is a wonderful option, many patients would obviously prefer to use their own genetics if at all possible. We have tried many different approaches to overcoming this challenge. One of those we'd hoped would work, but has not so far, was injecting a patient's eggs (oocytes) with the mitochondria from the patient's own ovarian stem cells. The idea was to, in effect, super-charge and improve the eggs. We are still exploring the viability of this method, but we've yet to see convincing proof of its benefit.

Finally, there is the promise of stem-cell technology, which holds promise for new developments across the field of medicine. In the field of infertility treatment, important work is being done to take stem cells and allow them to differentiate potentially into new eggs or new oocytes that could be fertilized and lead to embryos. Such work has already been accomplished in a mouse model, but live births are still pending. Once this has been successfully accomplished in animal models, we may see it tested in humans.

❧ *STREAMLINED TREATMENTS ARE ON THE HORIZON*

Researchers in our field are not simply focusing on major breakthroughs. They are also looking at ways to streamline and simplify infertility treatment and I believe patients will benefit greatly from this over the next ten years. As every patient is well aware, IVF treatments are quite complicated, lengthy, and painful with multiple injections daily. We look forward to either treatments with fewer injections or possibly avoiding injections altogether with alternative ways to administer drugs. This would make a fertility treatment cycle easier on the patient.

Typically, patients in the past have viewed IVF as the last resort. They'll go through the basic tests and then try fertility drugs like clomiphene pills and direct sperm injection into the uterus, or IUI. If several of those treatments don't work, they'll move to hormone shots and more IUIs before trying IVF.

Recently, however, as doctors in our field have become more successful with IVF, an increasing number of patients are skipping those other methods and going right to IVF treatments. Studies have shown that women who go directly to IVF spend less money and less time in treatment and have more babies.

I think we will be seeing more and more patients do the basic tests and then moving quickly into IVF, rather than seeing IVF as the "last resort." This is already happening particularly in the case of older patients who don't want to spend either the time or money going through treatments that have lower success rates than IVF. For women under forty, the average pregnancy rate is about 40 percent in the first attempt and it can be higher at some clinics. For those forty to forty-two, the average pregnancy rate with IVF nationwide is about 15 percent and for those over forty-two, it is 5 percent or less.

Recent studies sponsored by the National Institutes of Health by researchers at Dartmouth and in Boston found that women who fast-tracked to IVF became pregnant three months faster on average and spent ten thousand dollars less than those who tried the more traditional treatments first. Those three months can make a huge difference in reducing stress, particularly for women in their late thirties and older. IVF also allows doctors to choose how many embryos to transfer, reducing the risks of multiple births.

I'm not saying that going right to IVF is the way every woman should go. This is a more invasive procedure and it uses stronger drugs that are more likely to result in ovarian overstimulation, and there is a minor surgery to retrieve the eggs. Yet, the benefits for taking the fast track to IVF are growing each year as methods are refined and the science advances.

If we can make this treatment simpler and less intrusive for the patient, while at the same time improving success rates and lowering multiple gestation rates, we will truly have advanced the field. With the galloping pace of technology, I am sure many of these challenges will be overcome and great leaps forward in fertility treatments will be realized over the next ten years. I encourage you to stay on top of developments by monitoring the news pages on our website and others that are on the cutting edge of reproductive science.

ABOUT THE AUTHOR

 Doctor William Schoolcraft is the founder and medical director of the Colorado Center for Reproductive Medicine, one of the most renowned clinics in the nation. In the last thirty-plus years, Doctor Schoolcraft and his colleagues have helped families conceive more than 22,000 in vitro fertilization (IVF) babies. CCRM has pioneered significant research breakthroughs in IVF including in vitro culture of human embryos to the blastocyst stage, novel protocols for the treatment of poor responders, and blastocyst comprehensive chromosome screening.

Doctor Schoolcraft completed his medical training at the University of Kansas and finished his residency in obstetrics and gynecology in 1983 at UCLA. After his fellowship at UCLA under the direction of Doctor David Meldrum, one of the pioneers in the field of IVF, Doctor Schoolcraft established CCRM. Doctor Schoolcraft and CCRM have developed a national reputation for excellence in both pregnancy and delivery rates.

Considered a leading authority on fertility, Doctor Schoolcraft has been interviewed by *The New York Times, Wall Street Journal, Denver Post, Men's Health,* Today.com, and has appeared as a fertility expert on numerous television programs, including *Good Morning America, The Doctors,* and multiple Colorado news programs. Doctor Schoolcraft is the author of the book *If At First You Don't Conceive: A Complete Guide to Infertility from One of the Nation's Leading Clinics.*

Doctor Schoolcraft has authored countless international peer-reviewed scientific articles, which have documented the Center's high success rates including novel methods to stimulate multiple follicular growth, blastocyst comprehensive chromosomal screening, single blastocyst transfers, novel protocols for the treatment of poor responders, and vitrification methods for oocyte and embryo cryopreservation.

For twenty-two years in a row, *5280 Magazine* has named Dr. Schoolcraft a "Top Doctor" in the state of Colorado. Doctor Schoolcraft is also one of a handful of physicians that is Board Certified by the American Board of Bioanalysis as a high complexity clinical laboratory director.

Doctor Schoolcraft has also been named Castle Connolly 2017 Top Doctor in Reproductive Endocrinology/Infertility.